MACHINE LEARNING AND PATTERN RECOGNITION METHODS IN CHEMISTRY FROM MULTIVARIATE AND DATA DRIVEN MODELING

MACHINE LEARNING AND PATTERN RECOGNITION METHODS IN CHEMISTRY FROM MULTIVARIATE AND DATA DRIVEN MODELING

Edited by

JAHAN B. GHASEMI

Faculty of Chemistry, University of Tehran, Tehran, Iran

ELSEVIER

Elsevier
Radarweg 29, PO Box 211, 1000 AE Amsterdam, Netherlands
The Boulevard, Langford Lane, Kidlington, Oxford OX5 1GB, United Kingdom
50 Hampshire Street, 5th Floor, Cambridge, MA 02139, United States

ISBN: 978-0-323-90408-7

For information on all Elsevier publications
visit our website at https://www.elsevier.com/books-and-journals

Publisher: Susan Dennis
Acquisitions Editor: Charles Bath
Editorial Project Manager: Catherine Costello
Production Project Manager: R.Vijay Bharath
Cover Designer: Matthew Limbert

Typeset by STRAIVE, India

Contents

Contributors

Karim Abbasi
Laboratory of System Biology, Bioinformatics & Artificial Intelligent in Medicine (LBB&AI), Faculty of Mathematics and Computer Science, Kharazmi University, Tehran, Iran

Hamid Abdollahi
Department of Chemistry, Institute for Advanced Studies in Basic Sciences (IASBS), Zanjan, Iran

José Manuel Amigo
Department of Analytical Chemistry, University of the Basque Country, Leioa; IKER-BASQUE, Basque Foundation for Science, Bilbao, Spain

Alessandra Biancolillo
Department of Physical and Chemical Sciences, University of L'Aquila, Coppito, Italy

Paul J. Gemperline
East Carolina University, Greenville, NC, United States

Jahan B. Ghasemi
Faculty of Chemistry, University of Tehran, Tehran, Iran

Reaha Goyetche
Department of Analytical Chemistry, University of the Basque Country, Leioa, Spain

Mohammad Hossein Karimi-Jafari
Department of Bioinformatics, Institute of Biochemistry and Biophysics, University of Tehran, Tehran, Iran

Somaiyeh Khodadadi Karimvand
Department of Chemistry, Institute for Advanced Studies in Basic Sciences, Zanjan, Iran

Leire Kortazar
Department of Analytical Chemistry, University of the Basque Country, Leioa; Plentzia Marine Station, University of the Basque Country (PiE-UPV/EHU), Plentzia, Spain

Mahsa AKbari Lakeh
Department of Chemistry, Institute for Advanced Studies in Basic Sciences, Zanjan, Iran

Federico Marini
Department of Chemistry, University of Rome La Sapienza, Rome, Italy

Mohsen Namazi
Department of Bioinformatics, Institute of Biochemistry and Biophysics, University of Tehran, Tehran, Iran

Farzad Qassemi
AIQTech Inc, Toronto, ON, Canada

Parvin Razzaghi
Department of Computer Science and Information Technology, Institute for Advanced Studies in Basic Sciences (IASBS), Zanjan, Iran

Somaye Vali Zade
Halal Research Center of IRI, Food and Drug Administration, Ministry of Health and Medical Education, Tehran, Iran

Preface

Human has always been looking for a way to do two things: to count and to measure. Humanity owes the knowledge of the definitions of coordinates, distance, and, in principle, the very existence of quantitative measurements in the sciences to the invaluable efforts of the French mathematician and philosopher René Descartes. In most cases, measuring requires a tool that can be used to characterize an object or thing. At first, an object is stimulated by a variable and then its response is recorded. In other words, all beings have a minimum of characteristics that are either common or unique to others. That is to say, measurement aims to see and recognize differences and similarities between individuals. Observations and measurements create a large amount of data designated to each member of a statistical population. Now, by projecting these features in a coordinate system, we create a space where each target population member occupies a specific position. Let us not forget that objects in such a space can be close or far apart. Their distinction or difference is essentially defined in terms of distances. Depending on the nature of the discipline, for instance whether it is natural or social sciences, a study demands a particular interpretation of space, distance, and similarity in regard to the members, objects, or individuals of the target community. In experimental sciences such as chemistry, measurements and observations, and any monitoring and investigation of objects in space, are to uncover the existence of buried relationships, patterns, and distribution and to find a quantitative mathematical relation that relates the attributes and properties of objects to responses or sometimes their characteristics. The former leads to subjects such as pattern recognition, classification, and clustering, and the latter results in concepts such as regression and calibration.

Machine learning methods and artificial intelligence can perhaps be traced back to the work of the eminent English scientist Alan Turing. Since then, tremendous advances in the development of new and powerful techniques in all scientific and technical disciplines have led to the provision of tools that have contributed significantly to the correct understanding of many phenomena.

The use of machine learning and artificial intelligence methods in solving chemical problems to measure multiple species, despite the considerable overlap in the response and the identification of new compounds, is inevitable. Considering the explosive growth of low-cost devices and the large amount of big data they produce, multivariate machine learning methods play a crucial role in rapidly processing and extracting useful information from various numerical, textual, and even image data types. Deep learning and neural networks are powerful tools in the hands of chemists to solve many of the problems they face.

In this book, we have tried to show concisely but precisely the role of these methods in solving chemical problems.

Jahan B. Ghasemi
Tehran, November 2022

CHAPTER 1

Soft constraints in curve resolution problems

Paul J. Gemperline[a] and Mahsa AKbari Lakeh[b]
[a]East Carolina University, Greenville, NC, United States
[b]Department of Chemistry, Institute for Advanced Studies in Basic Sciences, Zanjan, Iran

1. Introduction

Self-modeling curve resolution (SMCR) (Lawton & Sylvestre, 1971) describes a set of powerful tools for estimating the pure component spectra and composition profiles from the spectroscopic data matrices. The only requirement of this mathematical resolution is the bilinearity of the chemical data, meaning that the total response at one particular wavelength should be the sum over all the contributions of each spectroscopically active constituent according to Beer-Lambert law. At a constant optical path length, the Beer-Lambert law can be described by Eq. (1.1):

$$A_\lambda = s_{\lambda 1}c_1 + s_{\lambda 2}c_2 + \cdots + s_{\lambda m}c_m = \sum_{j=1}^{m} s_{\lambda j}c_j \tag{1.1}$$

where A_λ is the absorbance at wavelength λ, s_λ is the molar absorptivity of the active chemical components in mol L^{-1} cm^{-1}, and c is the concentration in mol L^{-1}. Eq. (1.1) can be written as the absorbance at each wavelength for every sample if complete spectra are taken over several samples. This system of equations allows the application of a matrix notation shown in Eq. (1.2):

$$\mathbf{A} = \mathbf{C}\,\mathbf{S}^{\mathrm{T}} + \mathbf{E} \tag{1.2}$$

where **A** ($ns \times n\lambda$) is the matrix of absorbance that consists of absorption spectra measured at $n\lambda$ wavelengths from ns samples. Columns of matrix **C** ($ns \times nc$) contain the pure concentration profiles, and rows of matrix **S** contain the pure spectra of the absorbing species. **E** ($ns \times n\lambda$) is the matrix of residuals and contains the measurement errors.

The SMCR research hypothesis. In typical SMCR applications, several constraints are used to refine the initial estimates of concentration or spectral profiles to the final solution. These constraints are chosen based on the available information from the chemical system under study, such as

Machine Learning and Pattern Recognition Methods in Chemistry from Multivariate and Data Driven Modeling
https://doi.org/10.1016/B978-0-323-90408-7.00004-6

non-negativity, unimodality, equality, known-value constraints, etc. There is an implied research hypothesis in SMCR applications (Gemperline & Cash, 2003) that should be recognized and considered carefully: "There exists an *unconstrained* bilinear model with unimodal, non-negative pure component concentration profiles and pure component non-negative spectra that fits the data matrix of measurements obtained from the evolving system (Gemperline & Cash, 2003)." Normally, this hypothesis is examined through an iterative process of applying hard constraints, i.e., strictly forced constraints, to fit the data until convergence is achieved. In cases where the SMCR research hypothesis is true, the final model contains no active constraints. However, in many cases, due to noise and nonideal behaviors of chemical and/or spectroscopic response, active constraints are present in the resultant SMCR model. The detection of active constraints is important both from a theoretical and from a practical point of view. Tests for active constraints can be simply done by turning the constraints off at the solution point and then comparing the constrained fit with the unconstrained one. In the case of strong active constraints, when the difference between the two fits is greater than what might be expected from noise, a significant deviation from the original hypothesis may occur. The research hypothesis is rejected if many strong active constraints are present at the final solution. It is worth noting that in most MCR fits, instead of producing a unique solution, these methods produce a range of feasible solutions, which fulfill the constraints. Therefore, if one gets a small lack-of-fit with no strong active constraints, one cannot prove that the hypothesized model is correct, as in testing all scientific hypotheses.

Most often, the first step in fitting SMCR models is to select some initial starting point followed by iterative refinement while considering physicochemical limitations in the form of constraints until convergence to a meaningful solution is achieved. The starting point or initial "guess" can be selected using different methods, including the "needle search" method (Gemperline, 1984, 1986) or evolving factor analysis (EFA) (Gampp, Maeder, Meyer, & Zuberbuehler, 1987; Maeder, 1987; Maeder & Zuberbuehler, 1986), when good information is available about the concentration profiles, or variable selection methods to approximate the pure component spectra or pure variables. The method of constrained alternating least squares is used to iteratively refine the initial "unconstrained" estimates to yield better "constrained" solutions. In solving SMCR problems, the constraints are applied during the fitting process to ensure a smooth and monotonic convergence of the initial solution to the final result. The iterative refinement process by the application of alternating least-squares steps is repeated until no further improvement is seen in the estimated concentration or spectral profiles or when the maximum number of iterations is reached.

Normally, the spectroscopic techniques that are employed do not give a selective response for each chemical species present in a mixture, and the chemical separation techniques (for example, chromatography) cannot isolate one or more species from the mixture in a pure state; thus, the applied constraints do not add sufficient information to completely resolve the curve resolution problem. Thus, in most MCR fits, nonunique solutions are obtained. In such cases, it is recommended to calculate the whole range of feasible solutions for both identification and quantitation purposes. There are different SMCR techniques for calculating the range of feasible solutions either by determining a useful approximation or by defining exact ranges. The interested reader is referred to a good review on this subject (Golshan, Maeder, & Abdollahi, 2013). However, in some circumstances, unique solutions can be obtained in SMCR calculations under the implied constraints. The conditions that are required for getting unique solutions have been the subject of several papers (Karimvand et al., 2020; Lakeh, Abdollahi, & Rajkó, 2022; Rajkó, Abdollahi, Beyramysoltan, & Omidikia, 2015).

Hard versus soft constraints. One of the important factors that can affect the reliability of the MCR solutions is how the constraints are applied. Generally, there are two ways for introducing prior physicochemical information to the MCR model: hard and soft constraints. "Hard" constraints are strictly imposed on the model, while "soft" constraints accept small deviations from the previously known information. For example, small negative deviations are permitted when applying soft non-negative constraints. In practice, soft constraints are useful to allow for the effects of noise and non-ideal behaviors of chemical mixtures and responses. Several published studies illustrate that applying soft constraints in SMCR models usually causes fewer active constraints and lower model lack-of-fit in comparison with hard constraints. Thus, soft constraints are more likely to fit the original research hypothesis discussed before. Several algorithms have been proposed for implementing soft constraints in SMCR methods and also for studying the effect of these constraints on the range of feasible solutions. The main algorithms for imposing soft constraints are penalty alternating least squares (P-ALS) (Gemperline & Cash, 2003; Richards, Miller, & Gemperline, 2008), polygon inflation (Rahimdoust Mojdehi, Sawall, Neymeyr, & Abdollahi, 2016; Sawall et al., 2015), and grid search algorithm (Lakeh, Abdollahi, & Gemperline, 2020; Tavakkoli, Abdollahi, & Gemperline, 2020). The constraints that are implemented softly so far are non-negativity, equality, closure, unimodality, known-value, windowing, and trilinearity. In the remaining sections of this chapter, we describe each different algorithm of applying soft constraints in detail, including how these algorithms can be used to construct the above-mentioned constraints. After presenting the details of algorithms, we present the effect of soft constraints on the range of feasible solutions using several simulated and experimental data sets.

2. Basic concepts and theory

2.1 Penalty alternating least-squares (P-ALS) algorithm

The P-ALS algorithm (Gemperline & Cash, 2003) was developed as a strategy to minimize the effects of noise and nonideal factors by enabling user control in the implementation of constraints. To achieve this goal, P-ALS follows the same steps discussed before to alternatively estimate the concentration profiles, **C**, and pure spectra, **S**, of the components; however, it applies a user-defined least-squares optimization approach to implement the constraints. The P-ALS algorithm can be summarized in the following steps as is shown in Fig. 1.1.

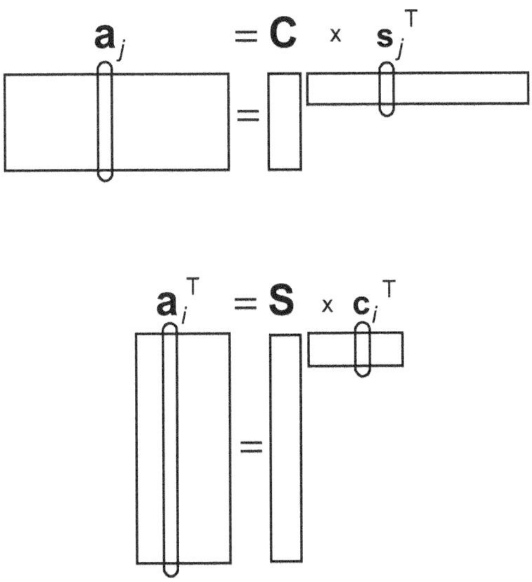

Fig. 1.1 Schematic representation of P-ALS algorithm for (top) finding the rows of the spectral matrix, **S**, and (bottom) finding the rows of the concentration matrix, **C**. *(From Gemperline, P. J. & Cash, E. (2003). Advantages of soft versus hard constraints in self-modeling curve resolution problems. Alternating least squares with penalty functions.* Analytical Chemistry, *75(16), 4236–4243.)*

1. An initial estimate of concentration profiles, $\mathbf{C}_0$, or pure spectra, $\mathbf{S}_0$, is selected by any number of popular methods.
2. Row-wise least-squares penalty functions are used to fit the initial unconstrained $\mathbf{C}_0$ or $\mathbf{S}_0$ producing better "constrained" estimates.
   a. Given some initial or intermediate estimate of $\mathbf{C}$, find every row of $\mathbf{S}$ (or every column of $\mathbf{S}^{\mathrm{T}}$) such that it minimizes $\|\mathbf{a}_{:,j} - \mathbf{C}\mathbf{s}_{:,j}^{\mathrm{T}}\|$ based on the constraints imposed on that row such as $\mathbf{s}_{:,j}^{\mathrm{T}} = \mathbf{g}_{:,j}$, where $\mathbf{g}_{:,j}$ specifies the constraint goals for spectra (more about this later). Based on this procedure, each row of $\mathbf{S}$ is estimated individually using least-squares penalty functions.
   b. Given some initial or intermediate estimate of $\mathbf{S}^{\mathrm{T}}$, find every row of $\mathbf{C}$ (or every column of $\mathbf{C}^{\mathrm{T}}$) such that it minimizes $\|\mathbf{a}_{i,:}^{\mathrm{T}} - \mathbf{S}\mathbf{c}_{i,:}^{\mathrm{T}}\|$ based on the constraints imposed on that row such as $\mathbf{c}_{i,:} = \mathbf{g}_{i,:}$, where $\mathbf{g}_{:,j}$ specifies the constraint goals for concentration profiles.
3. The constrained alternating least-squares steps are repeated iteratively until no further refinement in the estimated concentration and pure spectra can be observed or maximum number of iterations is achieved.
   - Equality constraints

Soft equality constraints can be applied for both row-wise P-ALS least-squares problems (2a and 2b) using penalty functions. The P-ALS subproblems can be generalized to the form of $\mathbf{y} = \mathbf{X}\mathbf{b}$. The least-squares problem can be solved for $\mathbf{b}$ using the equality-constrained weighting method (Van Benthem, Keenan & Haaland, 2002) by augmenting $\mathbf{y}$ with the vector of constrained values ($\mathbf{g}$) and $\mathbf{X}$ with a constraint matrix ($\mathbf{H}$) consisting of only one and zero values. The zero elements in the constraint matrix specify which species are not constrained. The model is shown in Eq. (1.3).

$$\begin{bmatrix} \mathbf{y} \\ \lambda\mathbf{g} \end{bmatrix} \cong \begin{bmatrix} \mathbf{X} \\ \lambda\mathbf{H} \end{bmatrix} \mathbf{b} \tag{1.3}$$

When imposing equality constraints with desired $\mathbf{g}$ values of spectral profiles, for every j, $\mathbf{y}$ would be the jth column of data ($\mathbf{a}_{:,j}$), $\mathbf{X}$ would be the initial or intermediate estimate of $\mathbf{C}$, and $\mathbf{b}$ would be the jth row of spectral profiles ($\mathbf{s}_{:,j}\mathrm{T}$). If reference data is available for concentrations and $\mathbf{g}$ contains the desired values, for every i, $\mathbf{y}$ would be the ith row of transposed data ($\mathbf{a}_{i,:}^{\mathrm{T}}$), $\mathbf{X}$ would be the initial or intermediate estimate of $\mathbf{S}$, and $\mathbf{b}$ would be the ith row of the transposed concentration profiles ($\mathbf{c}_{i,:}^{\mathrm{T}}$).

In Eq. (1.3), λ is the penalty function weighting factor that gives flexibility in the implementation of constraints. It can be adjusted to small values relative to the size and measurement scale of the data set when soft constraints are desired, for example, in the case where a significant amount of noise is expected to propagate into the estimated responses. In the case that the impact of noise and nonideal behavior of responses is negligible on the measurements, λ can be set to large values for applying hard constraints. To reasonably weigh different problems, the penalty value can be tuned relative to the norm of $\mathbf{X}$, e.g., λ can be set to $0.01 \times$ norm ($\mathbf{X}$) for applying soft constraints or to $10 \times$ norm ($\mathbf{X}$) if hard constraints are desired.

- Non-negativity constraints

Soft non-negativity constraints can be applied by a simple modification of steps 2a and 2b of P-ALS algorithm. For imposing soft non-negativity constraints, in the first step, the unconstrained least-squares problem is solved, thus resulting in the coefficients vector ($\mathbf{b}$). For any negative coefficients, b_i, an equality constraint is set ($b_i = 0$). These equality constraints are imposed using the penalty functions described in the previous section. The penalty value can be tuned for applying hard or soft non-negativity constraints. If hard constraints are desired, large penalty weights should be used to force the P-ALS solution to the result that would be obtained from NNLS (Lawson & Hanson, 1995) or FNNLS (Bro & De Jong, 1997; Bro & Sidiropoulos, 1998). An example is given below to illustrate the implementation of non-negativity constraints by P-ALS algorithm.

$$\mathbf{y} = \begin{bmatrix} 0.3111 \\ 0.9234 \\ 0.4302 \\ 0.1848 \end{bmatrix} \quad \mathbf{X} = \begin{bmatrix} 0.1656 & 0.6892 & 0.2290 \\ 0.6020 & 0.7482 & 0.9133 \\ 0.2630 & 0.4505 & 0.1524 \\ 0.6541 & 0.0838 & 0.8258 \end{bmatrix} \quad \hat{\mathbf{b}} = \begin{bmatrix} -0.0827 \\ 0.6528 \\ 0.3708 \end{bmatrix}$$

Considering that the first coefficient is negative, $b_1 = -0.0827$, a soft non-negativity constraint can be imposed as shown below using penalty weighting function $\lambda = 1$ and equality constraints, giving $\mathbf{b}_{\text{P-ALS}(\lambda 1)} = [-0.0023,\ 0.6503,\ 0.3134]$. Harder non-negativity constraints can be applied using $\lambda = 10$, which gives the coefficient vector $\mathbf{b}_{\text{P-ALS}(\lambda 10)} = [-2 \times 10^{-5},\ 0.6503,\ 0.3117]$, and if hard non-negativity constraints are desired, a larger penalty weight $\lambda = 100$ can be used, which results in $\mathbf{b}_{\text{P-ALS}(\lambda:100)} = [-2 \times 10^{-7},\ 0.6503,\ 0.3117]$. It is worth noting that the solution of non-negative least-squares method (NNLS) $\mathbf{b}_{\text{NNLS}} = [0,\ 0.6503,\ 0.3117]$ compares favorably to the solution obtained with $\lambda = 100$. For other examples illustrating the implementation of non-negativity constraints with the P-ALS algorithm, refer to the original paper (Gemperline & Cash, 2003).

$$\mathbf{y} = \begin{bmatrix} 0.3111 \\ 0.9234 \\ 0.4302 \\ 0.1848 \\ 0 \end{bmatrix} \mathbf{X} = \begin{bmatrix} 0.1656 & 0.6892 & 0.2290 \\ 0.6020 & 0.7482 & 0.9133 \\ 0.2630 & 0.4505 & 0.1524 \\ 0.6541 & 0.0838 & 0.8258 \\ 1 & 0 & 0 \end{bmatrix} \hat{\mathbf{b}} = \begin{bmatrix} -0.0827 \\ 0.6528 \\ 0.3708 \end{bmatrix}$$

- Closure constraints

Soft closure constraints can be implemented by the P-ALS algorithm using least-squares penalty functions and equality constraints. Augmenting the initial or intermediate estimate of $\mathbf{S}^T$ with the constraint matrix, $\mathbf{H} = [1, 1, \ldots, 1]$, and $\mathbf{y}$ with the constrained value, $g = [1]$, and considering an adjustable penalty function, λ, give results with sum of all species concentration profiles equal to or approximately equal to one. For chemical reactions where a weighted sum of several species concentrations is constant, e.g., as in the following consecutive reaction ($2A \rightarrow B \rightarrow C$), the constraint matrix would be different (e.g., $\mathbf{H} = [1/2\ 1\ 1]$ and $g = [1]$ for the consecutive reaction). The previous example is used again to illustrate the implementation of closure constraints and also to illustrate the fact that different constraints can be applied in P-ALS in one step. Soft closure and non-negativity constraints are implemented simultaneously with a penalty weighting function $\lambda = 1$, giving the results shown below with $\text{sum}(\mathbf{b}_{\text{P-ALS}}) = 0.9799$.

$$\mathbf{y} = \begin{bmatrix} 0.3111 \\ 0.9234 \\ 0.4302 \\ 0.1848 \\ 0 \\ 1 \end{bmatrix} \mathbf{X} = \begin{bmatrix} 0.1656 & 0.6892 & 0.2290 \\ 0.6020 & 0.7482 & 0.9133 \\ 0.2630 & 0.4505 & 0.1524 \\ 0.6541 & 0.0838 & 0.8258 \\ 1 & 0 & 0 \\ 1 & 1 & 1 \end{bmatrix} \hat{\mathbf{b}} = \begin{bmatrix} 0.0026 \\ 0.6622 \\ 0.3150 \end{bmatrix}$$

- Unimodality constraints

Soft unimodality constraints can also be applied by the P-ALS algorithm via penalty functions and equality constraints. In the first step, standard unconstrained least-squares problem is solved for all rows of $\mathbf{C}$. Each resulting concentration profile, $\mathbf{c}_{:,j}$, is checked to find the global peak maximum. Inspecting the concentration values in both forward and backward directions from the global peak maximums should confirm no secondary local maximums. Note, for example, that if $c_{i,j} < c_{i+1,j}$, in forward direction, and if $c_{i,\,j} < c_{i-1,\,j}$, in backward direction, local maximums are present. In such cases, unimodality constraints must be implemented by setting $g_{i+1,\,j} = c_{i,\,j}$ or $g_{i-1,\,j} = c_{i,\,j}$. For an example of illustrating the implementation of unimodality constraints in the P-ALS algorithm, refer to the original paper (Gemperline & Cash, 2003).

2.2 Multi-way penalty alternating least-squares (P-ALS) algorithm

Multi-way P-ALS or NWAY P-ALS was proposed as an extension to the P-ALS approach, incorporating all the advantages of that method, as well as the second-order advantage for species common to two or more experiments. The multi-way P-ALS procedure can be described as simply unfolding the multi-way array to a matrix and then performing the standard P-ALS approach. A detailed description of the multi-way P-ALS approach is given below.

A three-way array $\underline{\mathbf{A}}$ of the size $I \times J \times K$ can be matricized as a concatenated two-way array (or matrix) in three different directions, that is, a row-wise augmented matrix $\mathbf{A}_r$ $(JK \times I)$, a column-wise concatenated matrix $\mathbf{A}_c$ $(IK \times J)$, and a tube-wise augmented matrix $\mathbf{A}_t$ $(IJ \times K)$. Fig. 1.2 represents different ways of matricizing a three-way array. In typical spectroscopic applications, and also in this chapter, it is supposed that the rows of arrays represent the mixture spectra, the columns represent the variables (e.g., wavelengths), and the tubes represent different experiments. Considering this assumption, the column-wise augmentation of the measurement matrices keeps the variables mode intact, so the information in the variables mode would not be mixed up.

The bilinear decomposition of the column-wise augmented data, $\mathbf{A}_c$, would result in an augmented concentration matrix $\mathbf{C}_c$ $(IK \times N)$ and a spectral matrix $\mathbf{S}^T$ $(N \times J)$ considering N chemical spectroscopically active species present in data. In the multi-way P-ALS procedure, starting with initial estimates of $\mathbf{C}_{c0}$ or $\mathbf{S}_0$, row-wise least-squares penalty functions are used to fit the unconstrained species profiles alternatively to the constrained estimates. Steps 2a and 2b of the P-ALS algorithm can be modified for multi-way P-ALS analysis as shown below.

a. Given some initial or intermediate estimate of $\mathbf{C}_c$, find every row of $\mathbf{S}$ (or every column of $\mathbf{S}^T$) such that it minimizes $\|\mathbf{a}_{c:,\,j} - \mathbf{C}_c\mathbf{s}^T_{:,\,j}\|$ based on the constraints imposed on that row such as $\mathbf{s}^T_{:,\,j} = \mathbf{g}_{:,\,j}$, where $\mathbf{g}_{:,\,j}$ specifies the constraint goals for spectra.

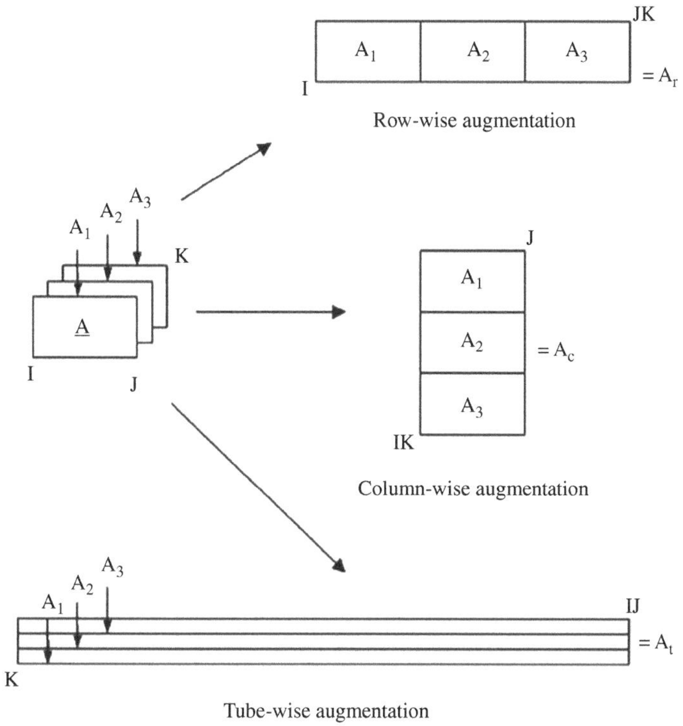

Fig. 1.2 Three different ways of augmentation of a three-way array (Richards et al., 2008).

b. Given some initial or intermediate estimate of $\mathbf{S}^T$, find every row of $\mathbf{C}_c$ (or every column of $\mathbf{C}_c^T$) such that it minimizes $\|\mathbf{a}^T_{ci,:} - \mathbf{S}\mathbf{c}^T_{ci,:}\|$ based on the constraints imposed on that row such as $\mathbf{c}_{ci,:} = \mathbf{g}_{i,:}$, where $\mathbf{g}_{:,j}$ specifies the constraint goals for concentration profiles.

These two steps are repeated iteratively to constrain $\mathbf{S}$ and $\mathbf{C}_c$ alternatively until convergence to the desired result. Fig. 1.3 illustrates the steps of applying constraints in the multi-way P-ALS method. The least-squares problem is solved using the adaptation of the constrained weighting method (Van Benthem et al., 2002) as described before.

An example is given to illustrate the implementation of the non-negativity and equality constraints simultaneously on the spectral values of a column-wise augmented data using the multi-way P-ALS approach. At first, the unconstrained least-squares problem is solved, and the resulting coefficient vector is illustrated below.

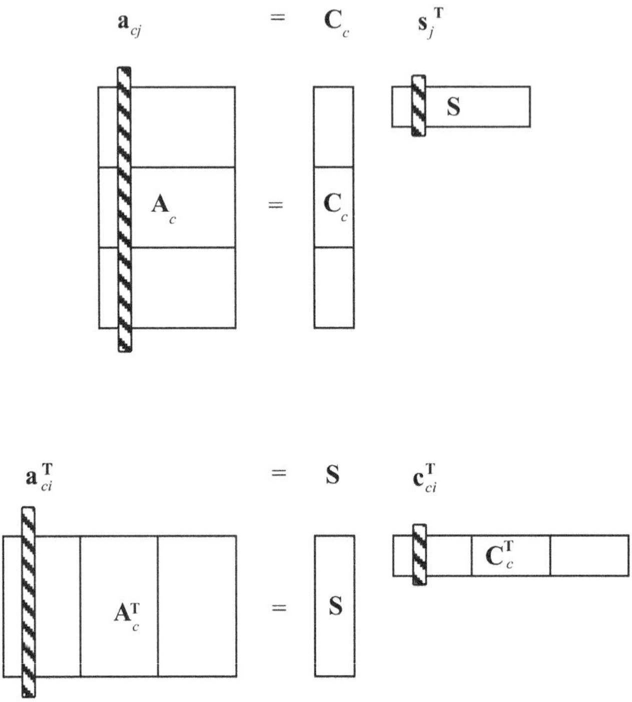

Fig. 1.3 Schematic representation of the multi-way P-ALS algorithm for (top) finding the rows of the spectral matrix, **S**, and (bottom) finding the rows of the concentration matrix, **C** (Richards et al., 2008).

$$y = \begin{bmatrix} 0.7094 \\ 0.7547 \\ 0.2760 \\ 0.6797 \\ \hline 0.6551 \\ 0.1626 \\ 0.1190 \\ 0.4984 \end{bmatrix} \quad \mathbf{X} = \begin{bmatrix} 0.8147 & 0.9575 & 0.4218 \\ 0.9058 & 0.9649 & 0.9157 \\ 0.1270 & 0.1576 & 0.7922 \\ 0.9134 & 0.9706 & 0.9595 \\ \hline 0.6324 & 0.9572 & 0.6557 \\ 0.0975 & 0.4854 & 0.0357 \\ 0.2785 & 0.8003 & 0.0357 \\ 0.5469 & 0.1419 & 0.9340 \end{bmatrix} \quad \widehat{\mathbf{b}} = \begin{bmatrix} 0.8184 \\ -0.0115 \\ 0.0421 \end{bmatrix} \tag{1.4}$$

In this example, **y** represents the first column of the augmented data $\mathbf{A}_c$, **X** represents the augmented concentration matrix $\mathbf{C}_c$, and **b** is the first column of $\mathbf{S}^T$. Suppose an equality constraint is required in the first coefficient

b_1, such that $b_1 = g_1 = 0.0500$. In addition, note that the second coefficient is negative ($b_2 = -0.0115$). So, both non-negativity and equality constraints were imposed iteratively using least-squares penalty functions with penalty weight $\lambda = 10$ giving $\mathbf{b}_{\text{P-ALS}(\lambda:10)} = [0.0107,\ 0.0059,\ 0.6112]$ in the first iteration. Softer constraints with smaller penalty would have resulted in smaller deviation from unconstrained values, meaning that more deviations are permitted from the non-negativity and equality constraints.

$$\mathbf{y} = \begin{bmatrix} 0.7094 \\ 0.7547 \\ 0.2760 \\ 0.6797 \\ \hline 0.6551 \\ 0.1626 \\ 0.1190 \\ 0.4984 \\ \hline 0.05 \\ 0 \end{bmatrix} \quad \mathbf{X} = \begin{bmatrix} 0.8147 & 0.9575 & 0.4218 \\ 0.9058 & 0.9649 & 0.9157 \\ 0.1270 & 0.1576 & 0.7922 \\ 0.9134 & 0.9706 & 0.9595 \\ \hline 0.6324 & 0.9572 & 0.6557 \\ 0.0975 & 0.4854 & 0.0357 \\ 0.2785 & 0.8003 & 0.8491 \\ 0.5469 & 0.1419 & 0.9340 \\ \hline 10 & 0 & 0 \\ 0 & 10 & 0 \end{bmatrix} \quad \hat{\mathbf{b}} = \begin{bmatrix} 0.0107 \\ 0.0059 \\ 0.6112 \end{bmatrix} \tag{1.5}$$

- Software implementation of P-ALS

GUIPRO is the name of a Matlab program that makes the implementation of P-ALS algorithm in a graphical user interface possible. This program has an NWAY P-ALS options window for applying different constraints on augmented matrices. The user can perform PCA, select the number of PCA factors, and perform SMCR on the spectroscopic data loaded from Matlab files or tab-separated data from EXCEL spreadsheets. Preprocessing can significantly improve the results of P-ALS analysis. Many different options are possible for doing preprocessing on the data, including selecting subranges for analysis, removing outliers detected by inspecting PCA scatter plots, correcting baseline shifts, and normalization of spectra. Before NWAY P-ALS analysis, the preprocessing should be done on each matrix of the augmented data individually, and the resultant matrices should be saved using the compatible pretreatment options. In the GUIPRO environment, up to 10 preprocessed individual data matrices (submatrices) can be loaded. To start the NWAY P-ALS algorithm, initial guesses of the augmented concentration profiles are needed. Several procedures can be applied for making the initial estimates, including the needle search technique or the evolving factor analysis (EFA) method. Using the GUIPRO graphical interface, different constraints, including non-negativity, unimodality, equality, and closure constraints, can be applied on the concentration profiles, and non-negativity and equality constraints can be applied on the spectral profiles. Separate penalty function weighting factors in the range of 0.01 to 20 can be selected for concentration and spectral profiles. For more detailed information about this program, please refer to Gemperline and Cash (2003) and Richards et al. (2008).

The P-ALS approach can demonstrate the impact of soft constraints on one set of feasible solutions of the curve resolution problem. However, a fundamentally different approach is to study the effect of soft constraints on the whole range of feasible solutions. The polygon inflation algorithm (Sawall, Kubis, Selent, Boerner, & Neymeyr, 2013; Sawall & Neymeyr, 2014) can be applied for this purpose in three-component systems by the use of a cost function that has the ability to apply soft constraints. The idea behind the polygon inflation algorithm is to approximate numerically each isolated area of feasible solutions (AFS) by a successive increase in the size of several predefined polygons. The feasibility of a certain concentration or spectral profile with the corresponding coordinates in the abstract space $(1, \alpha, \beta)$ is checked by solving a computationally expensive minimization problem. In this problem, a submatrix $\mathbf{P}$ (2×2) within $\mathbf{T}$ should be found to solve the factorization problem with the considered constraints.

$$\mathbf{T} = \begin{bmatrix} 1 & \alpha & \beta \\ 1 & p_{11} & p_{12} \\ 1 & p_{21} & p_{22} \end{bmatrix} \tag{1.6}$$

- Non-negativity constraints

The target function for soft non-negativity constraints is defined, as shown below, to accept different levels of deviations from non-negativity by changing the parameter ε (Rahimdoust Mojdehi et al., 2016).

$$\begin{aligned} f_{non-veg}(a, \beta, P) = & \sum_{i=1}^{3} \sum_{j=1}^{I} \min\left(0, \frac{C(j,\ i)}{\|C(:,\ i)\|_{\infty}} + \varepsilon\right)^2 \\ & + \sum_{i=1}^{3} \sum_{j=1}^{J} \min\left(0, \frac{S(j,\ i)}{\|S(i,\ :)\|_{\infty}} + \varepsilon\right)^2 \\ & + \|I_3 - T^{+}T\|_F^2 \end{aligned} \tag{1.7}$$

In the case where there are no active non-negativity constraints, $f_{non-neg}$ will equal zero; otherwise, its value indicates the amount of deviations from the non-negativity. A certain point (α, β) in the normalized abstract space (and its corresponding profile) is feasible if and only if $\min_{P} f_{non-neg}(\alpha, \beta, P)$ is zero or smaller than a threshold ε_f, e.g., $\varepsilon_f = 10^{-12}$. If additional soft constraints are desired, an extended cost function should be used.

- Unimodality constraints

Unimodality constraints are applied when the concentration profile of a chemical species has only one local minimum. In experimental conditions where measurement noise is present, small deviations from unimodality may be seen. A soft unimodality cost function can tolerate small descents or ascents against the main trend of the function using a control parameter $\omega \geq 0$. When all the columns of the estimated concentration matrix, **C**, are unimodal, the unimodality cost function, f_{uni}, will be equal to zero. In such cases, the control parameter can be set to zero. If the absolute value of the deviations from the unimodality constraints is larger than ω, the cost function will have a positive value equal to the sum of squares of all the deviation values. For more information about the unimodality cost function, please refer to Sawall et al. (2015).

- Closure constraints

Soft closure constraints can be applied in cases that small deviations from the principle of mass balance is acceptable, meaning that the sum of all rows of estimated **C** should be approximately constant. The deviations from the closure constraints are controlled using a parameter $\delta \geq 0$. The target function for closure constraints in the polygon inflation algorithm is defined as shown below (Sawall et al., 2015).

$$f_{clo}(a, \beta, P) = \sum_{i=1}^{I} \max\left(\left| 1 - C(i, \ :) \begin{pmatrix} x \\ y \\ z \end{pmatrix} \right| - \delta, 0 \right)^2 \tag{1.8}$$

$$\left\| \begin{pmatrix} 1 \\ \vdots \\ 1 \end{pmatrix} - C \begin{pmatrix} x \\ y \\ z \end{pmatrix} \right\|_2^2 \rightarrow \min \text{ with } x,y,z > 0 \tag{1.9}$$

where x, y, and z should be calculated as optimal scaling parameters of the columns of **C**. The cost function for closure f_{clo} results in a value of 0 if all rows of **C** are equal to a constant value.

- Equality constraints

The cost function for the equality constraints, $f_{equ}(\alpha, \beta, P)$, is defined directly from the constraint fulfillment as shown below. The range of the accepted deviations from the known-values, s_j^{given}, can be controlled using factor ϑ.

$$f_{equ}(a, \beta, P) = \sum_{j=1}^{J} \max\left(S_{ij} - \vartheta S_j^{given}\right)^2 \tag{1.10}$$

$$\vartheta = \frac{\sum_{i=1}^{j} s_j^{given} S_{ij}}{\sum_{j=1}^{J} \left(s_j^{given}\right)^2} \tag{1.11}$$

- Total cost function

When several soft constraints are desired to be applied, the total cost function can be defined as the weighted sum of all the objective functions that are needed. If the four constraints from the previous sections are considered, then the total cost function can be described as below:

$$\begin{aligned} f\,(\alpha, \beta, P) = & f_{non-neg}\,(\alpha, \beta, P, \varepsilon) + \gamma_{uni} f_{uni}\,(\alpha, \beta, P, \omega) \\ & + \gamma_{clo} f_{clo}\,(\alpha, \beta, P, \delta) + \gamma_{equ} f_{equ}\,(\alpha, \beta, P, \vartheta) \end{aligned} \tag{1.12}$$

where γ_{uni}, γ_{clo}, γ_{equ} are the weights that can vary between zero and one. As shown before, each pair of (α, β) defines a certain point of the AFS if $\min_P f(\alpha, \beta, P)$ is less than a control parameter.

$$f\,(\alpha, \beta, P) = \min_P f(\alpha, \beta, P) \leq \varepsilon_f \tag{1.13}$$

The value of the control parameter, $\varepsilon_f > 0$, should be chosen with caution especially for noisy experimental data. The control parameter for each constraint (ε, ω, δ, and ϑ) should be selected based on the acceptable amount of noise and deviations from the ideal responses. Details on the impact of the tunable parameters and the way to compute them are explained briefly in Sawall et al. (2015).

2.3 Grid search algorithm for applying soft constraints

The grid search method was proposed for numerically approximating the area of feasible solutions (AFSs) for 2-, 3-, and 4-component systems (Golshan, Abdollahi, & Maeder, 2011; Golshan et al., 2013; Vosough, Mason, Tauler, Jalali-Heravi, & Maeder, 2006). As the most informative part of the AFS is its boundary, which is a contiguous polygon, a simplex algorithm was employed to define the border and thus the complete AFS. The grid search algorithm can deal with noisy data sets and different restrictions coming from prior information of the system under study. In addition, this algorithm can be applied for studying the effect of various soft constraints on the range of feasible solutions. To our knowledge, only soft non-negativity, soft known-value, and soft trilinearity constraints have been employed using the grid search method.

- Soft non-negativity constraints

Soft non-negativity constraints were applied in the systematic grid search algorithm for calculating soft constrained AFSs (Lakeh et al., 2020). In this algorithm, the estimated concentration and spectral profiles were required to be greater than noise estimates $-\xi_c$ and $-\xi_s$, respectively.

$$\mathbf{c} > -\xi_c, \quad \mathbf{s} > -\xi_s \tag{1.14}$$

In the above equation, the application of ξ_c and ξ_s instead of zero allows small deviations from the non-negativity constraints in zero-concentration or zero-spectral regions of the resolved profiles. The noise estimates (ξ_c and ξ_s) were calculated for each concentration and spectral profile individually since the experimental noise propagates into the resolved profiles in different amounts. The first step to calculate the noise estimates was to smooth each individual profile in every iteration of the grid search algorithm. This means that not only the amount of noise propagation may be different from one profile to another, but it may also differ from one cycle of the grid search algorithm to another one. The smoothing was done using a binomial filter with the order equal to 5% of the profile length. In the next step, the difference between the unsmoothed and the smoothed profiles was smoothed again. A second residual was calculated by taking the difference between the unsmoothed residual and the smoothed residual. The second residual is assumed to contain only random deviations. The noise level of each individual profile, ξ_c or ξ_s, was estimated as 6 times the median absolute deviation (6 × MAD) of this second residual. The feasible transformation matrices, **T**, are those that produce estimated concentration and spectral values greater than the corresponding noise levels. The compliance of each species profile with the soft non-negativity constraints was examined in every iteration of the algorithm.

$$\delta_c = \sum_{i=1}^{ns} \sum_{j=1}^{3} \left(c_{ij} < -\xi_{cj}\right)^2 \text{ and } \delta_s = \sum_{i=1}^{n\lambda} \sum_{j=1}^{3} \left(S_{ij} < -\xi_{s_j}\right)^2, \ \delta = \delta_c + \delta_s \tag{1.15}$$

The parameter δ is a control parameter than can be used to evaluate the fulfillment of the soft non-negativity constraints. In cases of no deviations from the non-negativity constraints, δ is equal to zero.

- Known-value constraints

Partial or incomplete information about the contribution or signal of a chemical species can be applied as known-value constraints in SMCR methods. For implementing soft known-value constraints in the grid search method (Lakeh et al., 2020), in each cycle of the algorithm, profiles from which prior information is available must be scaled according to the true values. The scaling step is important because the ratio of known-values, and not their intensity, is important when comparing the estimated and true values. In the next step, the percent relative error between the estimated and true values should be calculated as a criterion for accepting the estimated profile as a feasible solution.

$$\mathrm{RE} = 100 \times \sqrt{\frac{\sum (v_t - v_c)^2}{\sum v_t^2}} \tag{1.16}$$

In the above equation, RE is the percent relative error, v_t is the true value, and v_c are the calculated or estimated values in each cycle of the grid search algorithm. Feasible solutions are found at grid points where their percent relative error is less than a predefined constant, e.g., 1%. This predefined constant controls the range of the accepted deviations from the known-values.

- Trilinearity constraints

Soft trilinearity constraints were implemented in the grid search algorithm (Tavakkoli et al., 2020) for allowing small deviations in the shape and position of the peak profiles of the components of interest from sample to sample. Implementing soft trilinearity constraints in the grid search algorithm is described using the following procedure.

At each cycle of the grid search algorithm:

1. Fold the estimated concentration profile for the desired component(s) into a matrix,
2. Apply PCA on the obtained matrix,
3. Calculate $k = \frac{\sum_{j=2}^{nf} eig_j}{\sum_{j=1}^{nf} eig_j} \times 100$, where nf is the number of augmented matrices
4. If $k \le \alpha$, then
   $\delta_\alpha = 0$
   else
   $\mathbf{R} = \mathbf{C} - \hat{\mathbf{C}}$, where $\hat{\mathbf{C}} = \left|\mathbf{U}_c(:,2:nf) \times \mathbf{S}_c(2:nf,2:nf) \times \mathbf{V}_c^{\mathrm{T}}(:,2:nf)\right|$

   $$\delta_\alpha = \sum\sum\left(\mathbf{R} \times \mathbf{R}^{\mathrm{T}}\right)$$

   end

Based on the proposed procedure, the estimated concentration profile of the component of interest, $\mathbf{c}_{aug}$ ($ns.nf \times 1$), should be folded into a matrix, $\mathbf{C}(ns \times nf)$, where ns is the number of samples (rows) of each data matrix, and nf is the number of augmented data. In the next step, principal component analysis (PCA) was performed on the $\mathbf{C}$ matrix to evaluate the deviations of the estimated concentration profiles from the trilinearity constraints. Since the estimated concentration profiles of the different matrices will have the same shape in the ideal cases where trilinearity holds, the first singular value would be significantly larger than the others. In such cases, the value of k is expected to be very small or close to zero, and mostly describes the noise variation. The control parameter, α, is used to tolerate small deviations from the trilinearity constraints in finding the feasible solutions. The larger the parameter α, the greater the accepted deviations from the trilinearity constraints. In the grid search algorithm with soft trilinearity constraints, the feasible solution grid points are those where the calculated value of k for the desired component is less than α. Fig. 1.4 represents the implementation of soft trilinearity constraints graphically.

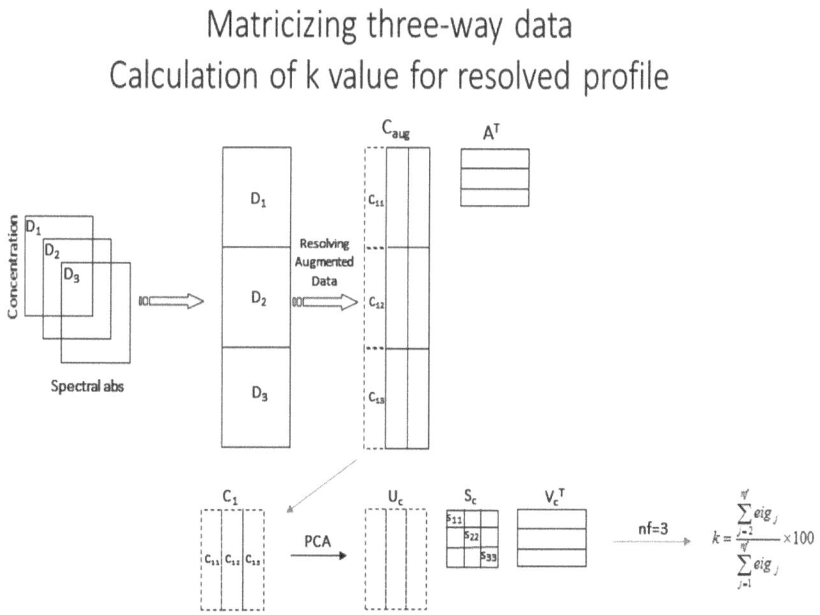

Fig. 1.4 Schematic representation of applying soft trilinearity constraints. The first column of the resolved concentration matrix is folded to give the matrix $\mathbf{C}_1$. The k value can be obtained by applying PCA on the $\mathbf{C}_1$, and calculating the eigen values. *(From Tavakkoli, E., Abdollahi, H., Gemperline, P.J. (2020). Soft-trilinear constraints for improved quantitation in multivariate curve resolution. Analyst, 145 (1), 223–232.)*

3. Applications

Gemperline and Cash used the P-ALS algorithm for comparing the results of applying hard and soft constraints in the SMCR analysis of four different examples (one simulated HPLC-DA data set and three experimental data sets) (Gemperline & Cash, 2003). They demonstrated that the results of the SMCR methods under hard constraints are distorted in many cases, and this can significantly affect the quantitative accuracy of the results. The average peak area errors for simulated HPLC-DA data with soft constraints using a penalty weight of 1.0 were less (~2%), whereas the errors with hard constraints, using a penalty weight of 20, were significantly larger (>12%). In the analysis of experimental data sets, a larger fraction of the total variance of data was modeled with soft constraints, and the resolved profiles under soft constraints were less noisy. In addition, applying soft constraints allowed a reasonable final model with improved estimated profiles in comparison with the hard constraints. Gemperline and coworkers (Lakeh et al., 2022) also demonstrated the impact of using soft constraints in the analysis of three-way data. The results revealed that the application of the NWAY P-ALS algorithm with soft constraints allows a reduction in the number of influential active constraints and enables monotonical and smooth convergence.

Neymeyr, Abdollahi, and their coworkers demonstrated the impact of soft constraints on the range of feasible solutions for three-component systems using FAC-PACK software (Sawall et al., 2013; Sawall & Neymeyr, 2014). They investigated the effect of several soft constraints on the AFSs, namely, constraints on non-negativity, equality, closure, unimodality, monotonicity, and windowing. Several simulated and experimental data sets were used for showing the advantages of applying soft versus hard constraints. In the case of non-negativity constraints, different levels of the control parameter, ε, were checked. Since larger ε values allow larger negative entries in the resolved profiles, the AFS segments were inflated by increasing the value of ε. In practice, real concentration or spectral profiles that construct the data matrix may show small deviations from non-negativity constraints. As it is shown in Rahimdoust Mojdehi et al. (2016), applying hard non-negativity constraints in such cases may define AFSs, which do not include the coordination of real noisy concentration or spectral profiles. This means that the accuracy of the feasible bands in the presence of noise is reduced. However, by applying the soft non-negativity constraints, the feasible regions could be expanded and the real known noisy profiles have the chance to be included in AFSs. Accordingly, the obtained results

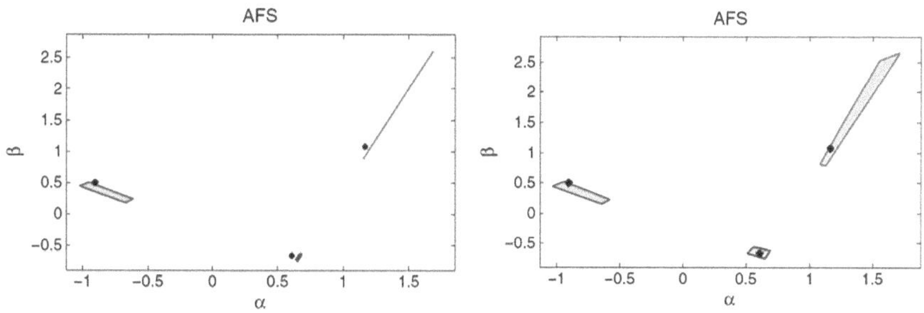

Fig. 1.5 The AFSs for data set 1 of the reference indicated below with (A) hard non-negativity constraints and (B) soft non-negativity constraints with $\varepsilon = 10^{-7}$. *(From Lakeh, M. A., Abdollahi, H., & Gemperline, P. J. (2020). Soft known-value constraints for improved quantitation in multivariate curve resolution.* Analytica Chimica Acta, 1105, *64–73.)*

demonstrated that the distortion in the estimated profiles due to noise and nonideal responses was reduced in the soft constrained AFSs. Fig. 1.5 depicts the effect of soft non-negativity constraints on the accuracy of the AFSs in the column space of a simulated data with normally distributed random noise (with a standard deviation of 0.001) described in Rahimdoust Mojdehi et al. (2016). The known concentration profiles of this data, which have small deviations from the non-negativity constraints, were projected in the abstract column space (indicated as black stars in Fig. 1.5). As can be seen in Fig. 1.5A, the AFSs obtained with hard constraints do not include the black stars (the coordinates of the known concentration profiles). This means that the boundaries of the hard constrained AFSs were not well defined and their corresponding profiles contained strong active constraints due to the presence of noise. However, as can be seen in Fig. 1.5B, applying soft non-negativity constraints with $\varepsilon = 10^{-7}$ caused a small expansion of the AFSs. In this way, profiles with small non-negative entries, as well as the known profiles, were accepted to be within the feasible regions.

Fig. 1.6 displays the results of the SMCR analysis of NIR batch reaction data, with the analysis described in Rahimdoust Mojdehi et al. (2016), and the original data from Gemperline and Cash (2003). In this data, NIR spectra were recorded during the acetylation of butanol with acetic anhydride in a jacketed flask at a thermostated temperature of 50°C. The reactor was initially charged with butanol, and then acetic anhydride was added to start the reaction with concentrated sulfuric acid as a catalyst. Spectra were recorded every 30 s in the range of 1100–2500 nm with 1-nm interval using a dispersive spectrometer and a fiber optic probe. The SMCR analysis of this data carried out

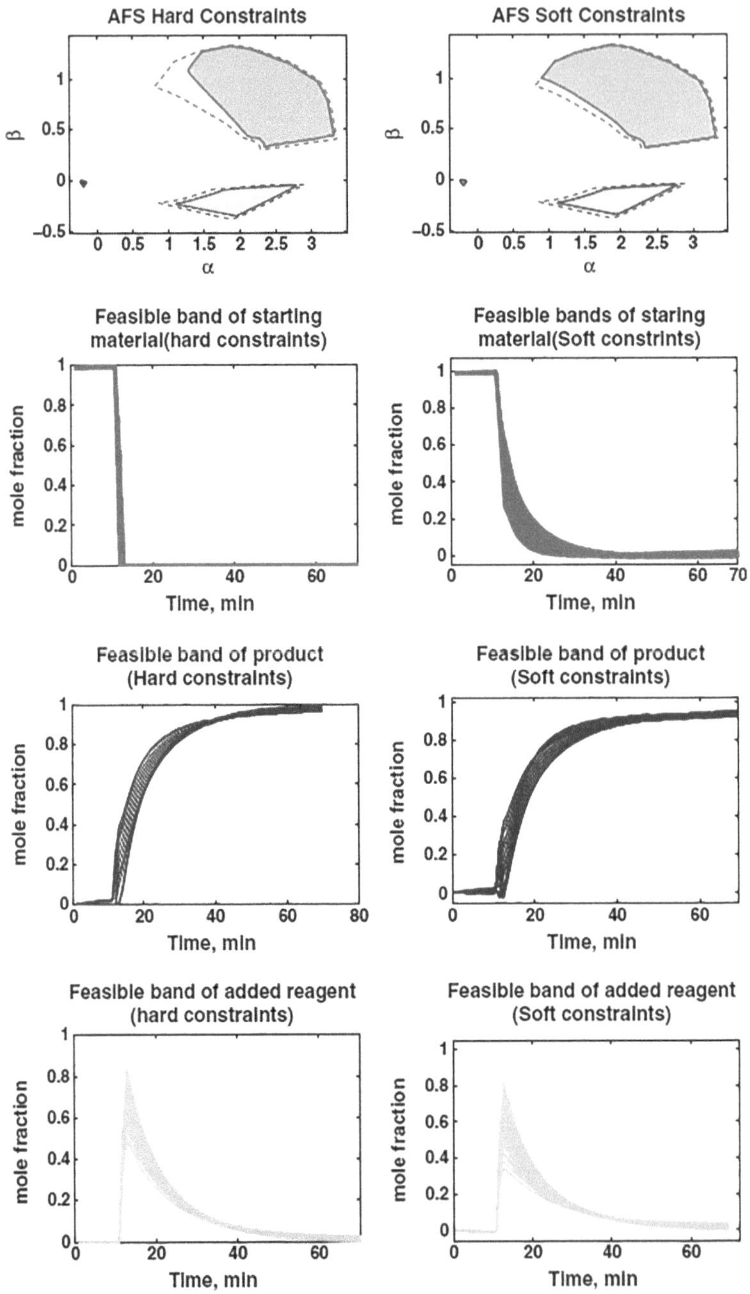

Fig. 1.6 Effect of hard and soft constraints on the AFSs and concentration bands of the NIR batch reaction data. Left: Results calculated by hard constraints. Right: Results calculated by soft constraints with $\varepsilon = 10^{-5}$, $\omega = 10^{-3}$, and $\delta = 10^{-1}$. *(From Rahimdoust Mojdehi, N., Sawall, M., Neymeyr, K., & Abdollahi, H. (2016). Investigating the effect of flexible constraints on the accuracy of self-modeling curve resolution methods in the presence of perturbations.* Journal of Chemometrics, 30*(5), 252–267.)*

under several constraints of non-negativity, unimodality, and closure. In the case of applying hard constraints, as can be seen in Fig. 1.6, the feasible bands of the starting material and the added reagent were estimated inaccurately. However, when soft constraints were used, the AFSs were calculated smoothly and their boundaries defined concentration bands with fewer active constraints. Furthermore, the feasible bands of the spectral profiles obtained with soft non-negativity, unimodality, and closure constraints contain the pure spectral profiles of acetic anhydride and butanol.

Additionally, it is demonstrated in Rahimdoust Mojdehi et al. (2016) that implementing soft constraints instead of hard constraints in data with no rotational ambiguity results in an increase in the accuracy of the model and a decrease in the model lack-of-fit.

Gemperline and his coworkers used the systematic grid search algorithm for implementing soft non-negativity, known-value, and trilinearity constraints (Lakeh et al., 2020; Tavakkoli et al., 2020). In one study (Lakeh et al., 2020), a second-order consecutive reaction was simulated, and four concentration values of the desired product were used to demonstrate the effect of implementing soft known-value constraints. It was assumed that the known concentration values were obtained from sampling of a batch reaction at four different times. Different levels of noise and deviations from the true values were considered to approach simulations of the experiments. The results of the grid search method obtained by imposing soft non-negativity constraints showed three distinct feasible regions for both concentration and spectral modes. As can be seen in Fig. 1.7, adding soft known-value constraints to the previous analysis caused a significant decrease in the amount of rotational ambiguity and, as a consequence, an increase in the accuracy of the results. When hard equality constraints were used in the curve resolution of this data, a distorted profile was obtained with its coordinates outside the AFSs in the column space. As the amount of noise and uncertainty of the known-values increased, the AFSs became larger, refer to Fig. 1.7C. The results of the analysis of experimental data are discussed below in more detail.

UV/visible batch reaction data. UV/visible spectra were recorded during the synthesis of acetylsalicylic acid (aspirin) in a glass reactor from the acetylation of salicylic acid (SA) with acetic anhydride (AA) in the presence of phosphoric acid as the catalyst. Spectra were recorded every 30 s using a UV/visible spectrograph that was equipped with an ATR fiber-optic probe. During the acetylation reaction, the desired product which was acetylsalicylic acid (ASA) formed an undesired by-product, acetylsalicylic anhydride (ASAA), due to

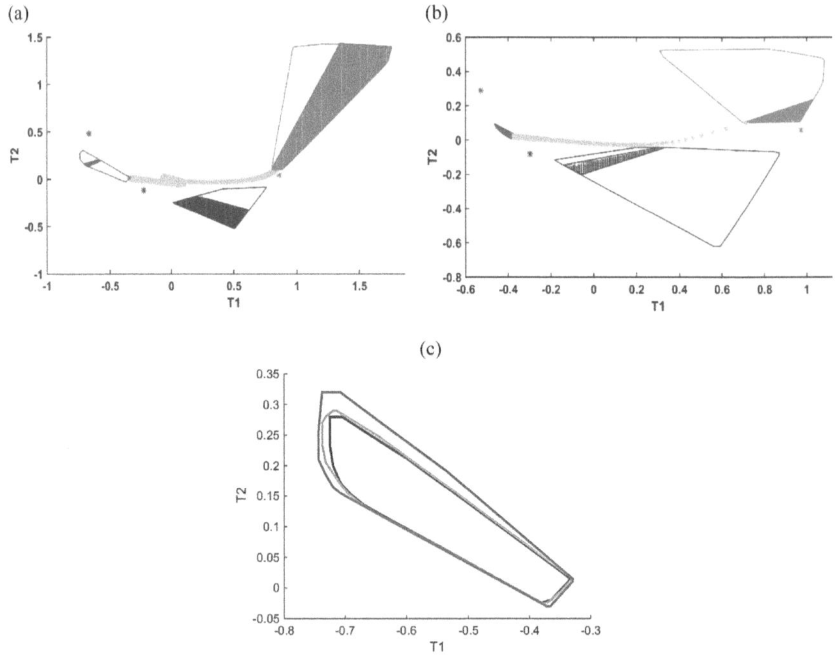

Fig. 1.7 The soft versus hard constraints. The AFSs of simulated data in (A) column space and (B) row space are shown by *solid lines*. The shaded parts, in (A) and (B), show the AFSs after the implementation of soft known-value constraints in addition to the non-negativity constraints. The *blue (light gray in the printed version), green (gray in the printed version)*, and *red (dark gray in the printed version) stars*, in (A) and (B), indicate the position of MCR-ALS results after applying hard non-negativity constraints. The AFS of the analyte in the column space is enlarged in (C). The *green (gray in the printed version), magenta (dark gray in the printed version)*, and *blue (light gray in the printed version) lines* show the AFS obtained by applying soft non-negativity constraint with the lowest, medium, and highest levels of noise, respectively. *(From Lakeh, M. A., Abdollahi, H., & Gemperline, P. J. (2020). Soft known-value constraints for improved quantitation in multivariate curve resolution.* Analytica Chimica Acta, 1105, *64–73.)*

the large excess of acetic anhydride (AA). To eliminate the by-product, deionized water was added to the mixture after 1 h from the start of the reaction. The ATR UV-Vis spectra were baseline-corrected using the minimum offset option of the GUIPRO package (Gemperline & Cash, 2003). The recorded spectra after 62 min, where the addition of water approximately eliminated the side product, were analyzed as a two-component mixture. Soft non-negativity constraints for both concentration and spectral values were implemented in the systematic grid search algorithm. The concentration band of

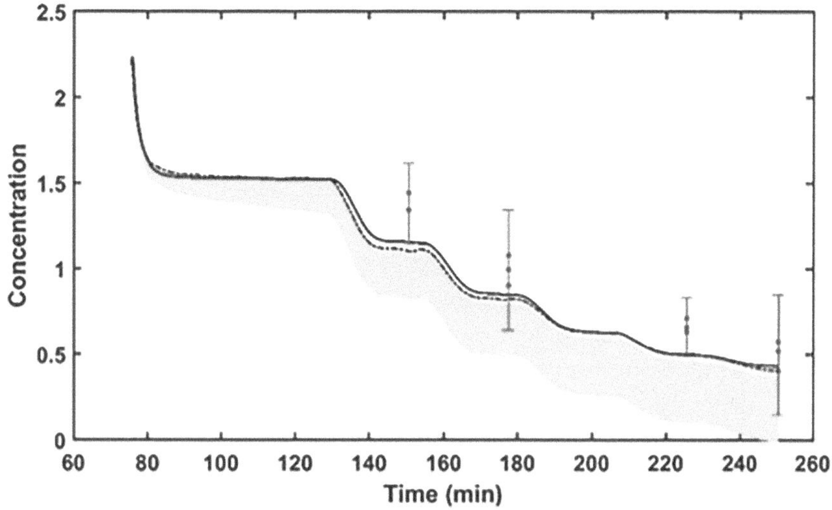

Fig. 1.8 The concentration band of the ASA after applying soft non-negativity constraints. The reference values and concentration intervals are shown as *red (gray in the printed version) dots* and *red (gray in the printed version) error bars*, respectively. None of the feasible solutions within the band fall into all of the concentration intervals. *(From Lakeh, M. A., Abdollahi, H., & Gemperline, P. J. (2020). Soft known-value constraints for improved quantitation in multivariate curve resolution.* Analytica Chimica Acta, 1105, *64–73.)*

ASA was calculated and is shown in Fig. 1.8. Every concentration profile within the band was normalized to the total concentration of SA. During the batch reaction, the concentration of ASA was measured at four different times, each time with three replicates, using an offline HPLC method. The concentration values obtained by replicates at each time point were used for calculating the confidence interval with $p = 0.99$, at that time. The individual concentration values of ASA and the confidence intervals are also shown in Fig. 1.8. As can be seen in Fig. 1.8, the concentration band of ASA did not fall into the concentration intervals. The deviations were so great that if the reference values were implemented as the soft known-value constraints, none of the band profiles would be accepted. This evidence suggests that the HPLC results were highly biased. Accordingly, calculating the area of feasible solutions gives deeper insights into the problem.

In another study, simulated and experimental data sets were used to evaluate the advantages of applying soft trilinearity constraints. In an augmented simulated data, the nonideal trilinear behavior of the analyte was modeled by shifting its peak maxima and also changing its peak width in different matrices. In this work, the impact of using different levels of trilinearity constraints was investigated on the range of feasible solutions of one of the data components (the analyte of interest). It was illustrated that by choosing smaller α values, the size of the analyte AFS and the width of the analyte concentration band decreased. As was shown in Tavakkoli et al. (2020), the estimation of the control parameter α is a crucial step in the implementation of soft trilinearity constraints as it determines the accuracy of the results. This parameter can be tuned based on prior knowledge about the chemical system under study. The results of applying hard trilinearity constraints on the analyte concentration profile showed the risk of obtaining a distorted profile and biased quantification. In addition, the PARAFAC2 model, which is known to give unique solutions in the analysis of non-trilinear data, was used for analyzing the simulated data. However, the solution of the PARAFAC2 model with non-negativity constraints was not within the range of feasible solutions, and as a result, the obtained concentration values were highly biased. This was because PARAFAC2 can only model certain types of non-trilinearity conditions, not all the possible ones, especially when there are changes in the peak widths. Similar conclusions were obtained in the analysis of the experimental data. For more detail, please refer to Tavakkoli et al. (2020).

4. Conclusions

By applying soft constraints in SMCR methods, small deviations from the constraint settings are allowed. Apart from the algorithm used, this approach may yield more accurate results in many more circumstances than those obtained by applying hard constraints. The estimated profiles of the SMCR methods with soft constraints are more likely to have fewer active constraints and distortions. These advantages can be attributed to the increased control and flexibility in the implementation of the constraints. Several software implementations are available to users for incorporating a variety of soft constraints, most of which are described in the text. Among them, those that calculate the whole range of feasible solutions under soft constraints are especially helpful. Soft constrained AFSs often have fewer active constraints and thus are more likely to fit the original research hypothesis in comparison with hard constrained AFS segments. The boundaries of AFS segments can be tuned to allow for acceptable levels of noise in a data set to propagate as small deviations into estimated profiles and estimate spectra that would otherwise not be allowed with hard constraints. In this chapter, the advantages of implementation of soft versus hard constraints are reviewed using several simulated and experimental data, which were published before. It has been demonstrated that in many cases, applying soft constraints, even for one of the components, can improve the accuracy of all the AFSs. Additionally, the approximation of the AFSs with soft constraints is often smooth and time efficient.

References

Bro, R., & De Jong, S. (1997). A fast non-negativity-constrained least squares algorithm. *Journal of Chemometrics: A Journal of the Chemometrics Society, 11*(5), 393–401.

Bro, R., & Sidiropoulos, N. D. (1998). Least squares algorithms under unimodality and non-negativity constraints. *Journal of Chemometrics: A Journal of the Chemometrics Society, 12*(4), 223–247.

Gampp, H., Maeder, M., Meyer, C. J., & Zuberbuehler, A. D. (1987). Evolving factor analysis. *Comments on Inorganic Chemistry, 6*(1), 41–60.

Gemperline, P. J. (1984). A priori estimates of the elution profiles of the pure components in overlapped liquid chromatography peaks using target factor analysis. *Journal of Chemical Information and Computer Sciences, 24*(4), 206–212.

Gemperline, P. J. (1986). Target transformation factor analysis with linear inequality constraints applied to spectroscopic-chromatographic data. *Analytical Chemistry, 58*(13), 2656–2663.

Gemperline, P. J., & Cash, E. (2003). Advantages of soft versus hard constraints in self-modeling curve resolution problems. Alternating least squares with penalty functions. *Analytical Chemistry, 75*(16), 4236–4243.

Golshan, A., Abdollahi, H., & Maeder, M. (2011). Resolution of rotational ambiguity for three-component systems. *Analytical Chemistry, 83*(3), 836–841.

Golshan, A., Maeder, M., & Abdollahi, H. (2013). Determination and visualization of rotational ambiguity in four-component systems. *Analytica Chimica Acta, 796*, 20–26.

Karimvand, S. K., Lakeh, M. A., Tavakkoli, E., Ghaffari, M., Omidikia, N., Abad, S. K. A., Rajkó, R., & Abdollahi, H. (2020). A general rule for uniqueness in self-modeling curve resolution methods. *Journal of Chemometrics, 34*(7), e3268.

Lakeh, M. A., Abdollahi, H., & Gemperline, P. J. (2020). Soft known-value constraints for improved quantitation in multivariate curve resolution. *Analytica Chimica Acta, 1105*, 64–73.

Lakeh, M. A., Abdollahi, H., & Rajkó, R. (2022). Predicting the uniqueness of single non-negative profiles estimated by multivariate curve resolution methods. *Analytica Chimica Acta, 1199*, 339575.

Lawson, C. L., & Hanson, R. J. (1995). *Solving least squares problems*. SIAM.

Lawton, W. H., & Sylvestre, E. A. (1971). Self modeling curve resolution. *Technometrics, 13*(3), 617–633.

Maeder, M. (1987). Evolving factor analysis for the resolution of overlapping chromatographic peaks. *Analytical Chemistry, 59*(3), 527–530.

Maeder, M., & Zuberbuehler, A. D. (1986). The resolution of overlapping chromatographic peaks by evolving factor analysis. *Analytica Chimica Acta, 181*, 287–291.

Rahimdoust Mojdehi, N., Sawall, M., Neymeyr, K., & Abdollahi, H. (2016). Investigating the effect of flexible constraints on the accuracy of self-modeling curve resolution methods in the presence of perturbations. *Journal of Chemometrics, 30*(5), 252–267.

Rajkó, R., Abdollahi, H., Beyramysoltan, S., & Omidikia, N. (2015). Definition and detection of data-based uniqueness in evaluating bilinear (two-way) chemical measurements. *Analytica Chimica Acta*, *855*, 21–33.

Richards, S., Miller, R., & Gemperline, P. (2008). Advantages of soft versus hard constraints in self-modeling curve resolution problems. Penalty alternating least squares (P-ALS) extension to multi-way problems. *Applied Spectroscopy*, *62*(2), 197–206.

Sawall, M., Kubis, C., Selent, D., Boerner, A., & Neymeyr, K. (2013). A fast polygon inflation algorithm to compute the area of feasible solutions for three-component systems. I: Concepts and applications. *Journal of Chemometrics*, *27*(5), 106–116.

Sawall, M., & Neymeyr, K. (2014). A fast polygon inflation algorithm to compute the area of feasible solutions for three-component systems. II: Theoretical foundation, inverse polygon inflation, and FAC-PACK implementation. *Journal of Chemometrics*, *28*(8), 633–644.

Sawall, M., Rahimdoust, N., Kubis, C., Schroeder, H., Selent, D., Hess, D., Abdollahi, H., Franke, R., Boerner, A., & Neymeyr, K. (2015). Soft constraints for reducing the intrinsic rotational ambiguity of the area of feasible solutions. *Chemometrics and Intelligent Laboratory Systems*, *149*, 140–150.

Tavakkoli, E., Abdollahi, H., & Gemperline, P. J. (2020). Soft-trilinear constraints for improved quantitation in multivariate curve resolution. *Analyst*, *145*(1), 223–232.

Van Benthem, M. H., Keenan, M. R., & Haaland, D. M. (2002). Application of equality constraints on variables during alternating least squares procedures. *Journal of Chemometrics: A Journal of the Chemometrics Society*, *16*(12), 613–622.

Vosough, M., Mason, C., Tauler, R., Jalali-Heravi, M., & Maeder, M. (2006). On rotational ambiguity in model-free analyses of multivariate data. *Journal of Chemometrics: A Journal of the Chemometrics Society*, *20*(6–7), 302–310.

CHAPTER 2

Multivariate predictive modeling and validation

Alessandra Biancolillo[a] **and Federico Marini**[b]
[a]Department of Physical and Chemical Sciences, University of L'Aquila, Coppito, Italy
[b]Department of Chemistry, University of Rome La Sapienza, Rome, Italy

Many problems in chemistry involve the prediction of one or more qualitative or quantitative properties based on the experimental data (Martens & Naes, 1991). Examples of such problems involve, for instance, the possibility of predicting protein or lipid content in food matrices based on NIR spectra or of diagnosing the onset of a disease through the MS or NMR analysis of serum samples. In the former case, the property to be predicted is of a quantitative nature, while in the latter, it is discrete (qualitative). This chapter presents the chemometric strategies most commonly used to formulate predictive models, i.e., models that relate one or more dependent variables Y (qualitative or quantitative) to a set of independent variables X.

Irrespective of its specific characteristics, building a predictive model requires not only that a relation between the X and the Y exists but also, operationally, that a reasonable mathematical formulation be postulated to express (or at least approximate) this relation:

$$y = f(\boldsymbol{x}) \tag{2.1}$$

Here, it should be stressed that since the exact form of the function $f(\boldsymbol{x})$ is unknown, it has then to be empirically estimated by making use of known data to approximate the functional relation in Eq. (2.1). Such estimation relies on the availability of a sufficient number of samples (which constitute the so-called training set), for which the values of both $\boldsymbol{x}$ and y are known. Operationally, model building is accomplished by defining an explicit mathematical formulation of the function $f(\boldsymbol{x})$, which is expressed in terms of some adjustable parameters (model coefficients), whose values are estimated based on the training data. While these concepts are equally valid for all kinds of predictive models, the techniques used for dealing with quantitative (regression) or qualitative (classification) responses present specificities

Machine Learning and Pattern Recognition Methods in Chemistry from Multivariate and Data Driven Modeling
https://doi.org/10.1016/B978-0-323-90408-7.00001-0

related to the nature of the dependent variable(s) and will be discussed individually in the remainder of this chapter.

1. Regression

Regression methods deal with modeling real-valued (quantitative) responses. In their simplest version, they postulate that the functional relationship between the response(s) and the independent variables (predictors) is a linear one. Assuming the case of a single dependent variable y and that $\boldsymbol{x}$ is a J-dimensional vector

$$\boldsymbol{x} = \begin{bmatrix} x_1 & x_2 & \cdots & x_j & \cdots & x_J \end{bmatrix} \tag{2.2}$$

such linear relationship can be expressed as:

$$y = \widehat{y} + e = b_0 + b_1\, x_1 + b_2\, x_2 + \cdots + + b_j\, x_j + \cdots + b_J\, x_J + e \tag{2.3}$$

where $\widehat{y}$ is the approximation (predicted value) of the response provided by the linear function, which is univocally determined by the values of the coefficients b_1, b_2, $\cdots$, b_J and of the intercept b_0. On the other hand, e indicates the residual, i.e., the difference between the true and predicted value of the response. Here, it should be stressed that if mean centering for both $\boldsymbol{x}$ and y is assumed, the intercept becomes zero, so in the remainder of the text, it will not be included in the models, i.e.,

$$\widehat{y} = b_1\, x_1 + b_2\, x_2 + \cdots + + b_j\, x_j + \cdots + b_J\, x_J \tag{2.4}$$

As preliminarily introduced above, model building corresponds to estimating the best values of the coefficients b_1, b_2, $\cdots$, b_J given the training data available, i.e., exploiting the fact that there is a set of observations for which not only the $\boldsymbol{x}$, but also the y is known. Since the aim of predictive modeling is to obtain accurate approximations of the dependent variable(s), in the majority of the cases, the definition of the optimal value of the coefficients involves some sort of minimization of the residuals (*least-squares* criterion).

Assuming that N training samples are available, the mathematical relationship described by Eqs. (2.3) and (2.4) should be valid for each of them and can be then generalized, in the case of the ith X-Y pair ($\boldsymbol{x}_i$, y_i) as:

$$y_i = \widehat{y}_i + e_i = b_1\, x_{i1} + b_2\, x_{i2} + \cdots + + b_j\, x_{ij} + \cdots + b_J\, x_{iJ} + e_i = \boldsymbol{x}_i^T\, \boldsymbol{b} + e_i \tag{2.5}$$

where $\boldsymbol{x}_i^T$ is a row vector made up of the values of the independent variables for the ith sample, while the model (regression) coefficients are collected in the column vector $\boldsymbol{b}$. By taking advantage of matrix notation, Eq. (2.5) for the whole training set can be written as:

$$\boldsymbol{y} = \widehat{\boldsymbol{y}} + \boldsymbol{e} = \boldsymbol{X}\boldsymbol{b} + \boldsymbol{e} \tag{2.6}$$

where the rows of the $\boldsymbol{X}$ matrix correspond to the experimental profiles of the different samples:

$$\boldsymbol{X} = \begin{bmatrix} \boldsymbol{x}_1^T \\ \boldsymbol{x}_2^T \\ \vdots \\ \boldsymbol{x}_i^T \\ \vdots \\ \boldsymbol{x}_N^T \end{bmatrix} \tag{2.7}$$

the true and predicted values of the response constitute the elements of the column vectors $\boldsymbol{y}$ and $\widehat{\boldsymbol{y}}$, respectively, while the residuals are collected in the vector $\boldsymbol{e}$.

1.1 Multiple linear regression (MLR)

When the ordinary least-squares approach is used to estimate the regression coefficients based on the training data, the corresponding technique is often referred to as multiple linear regression (MLR). In particular, the optimal value of the elements of $\boldsymbol{b}$ is sought as the one leading to the minimum approximation error on the training data (defined as the sum of squares of the residuals $E=\sum_{i=1}^{N} e_i^2$), as mathematically expressed by:

$$\underset{\boldsymbol{b}}{\operatorname{argmin}}\left\{\sum_{i=1}^{N} e_i^2\right\} = \underset{\boldsymbol{b}}{\operatorname{argmin}}\left\{\boldsymbol{e}^T\boldsymbol{e}\right\} = \underset{\boldsymbol{b}}{\operatorname{argmin}}\left\{(\boldsymbol{y}-\boldsymbol{X}\boldsymbol{b})^T(\boldsymbol{y}-\boldsymbol{X}\boldsymbol{b})\right\} \tag{2.8}$$

Solution to the condition expressed by Eq. (2.8) is obtained by differentiating E with respect to the elements of $\boldsymbol{b}$ and setting all the partial derivatives to zero. The corresponding values of the regression coefficients are:

$$\boldsymbol{b}_{MLR} = \left(\boldsymbol{X}^T\boldsymbol{X}\right)^{-1}\boldsymbol{X}^T\boldsymbol{y} \tag{2.9}$$

where the subscript MLR was added to highlight that the values of the coefficients for the linear model are those estimated with this technique.

Once the coefficients have been estimated using the training data, they can be used to predict the value of the response on any new sample, $\widehat{y}_{new}$, according to:

$$\widehat{y}_{new} = \boldsymbol{x}_{new}^T \boldsymbol{b} \tag{2.10}$$

where $\boldsymbol{x}_{new}^T$ is the row vector containing the experimental profile of the new sample.

So far, we have considered the case when only a single response has to be modeled; however, there can be cases where the number of dependent variables is higher than one and the approach described so far can be easily generalized to deal with problems involving a multivariate Y.

This is accomplished by expressing each of the responses as a linear function of the independent variables by means of its own set of regression coefficients:

$$\boldsymbol{y}_k = \widehat{\boldsymbol{y}}_k + \boldsymbol{e}_k = \boldsymbol{X}\,\boldsymbol{b}_k + \boldsymbol{e}_k \tag{2.11}$$

where the subscript k indicates that the model refers to the kth dependent variable (whose true and predicted values are collected in the vectors $\boldsymbol{y}_k$ and $\widehat{\boldsymbol{y}}_k$, respectively).

By gathering all the different predicted and true responses as columns of the matrices $\widehat{\boldsymbol{Y}}$ and $\boldsymbol{Y}$

$$\hat{\boldsymbol{Y}} = \begin{bmatrix} \hat{\boldsymbol{y}}_1 & \hat{\boldsymbol{y}}_2 & \cdots & \hat{\boldsymbol{y}}_k & \cdots & \hat{\boldsymbol{y}}_K \end{bmatrix} \tag{2.12}$$

$$\boldsymbol{Y} = \begin{bmatrix} \boldsymbol{y}_1 & \boldsymbol{y}_2 & \cdots & \boldsymbol{y}_k & \cdots & \boldsymbol{y}_K \end{bmatrix} \tag{2.13}$$

the residuals as columns of the matrix $\boldsymbol{E}$

$$\boldsymbol{E} = \begin{bmatrix} \boldsymbol{e}_1 & \boldsymbol{e}_2 & \cdots & \boldsymbol{e}_k & \cdots & \boldsymbol{e}_K \end{bmatrix} \tag{2.14}$$

and the regression coefficients in the matrix $\boldsymbol{B}$

$$\boldsymbol{B} = \begin{bmatrix} \boldsymbol{b}_1 & \boldsymbol{b}_2 & \cdots & \boldsymbol{b}_k & \cdots & \boldsymbol{b}_K \end{bmatrix} \tag{2.15}$$

Equation (2.11) can be globally rewritten for all the dependent variables as:

$$\boldsymbol{Y} = \widehat{\boldsymbol{Y}} + \boldsymbol{E} = \boldsymbol{X}\boldsymbol{B} + \boldsymbol{E} \tag{2.16}$$

Given the relation summarized by Eq. (2.16), the optimal solution, i.e., the best set of regression coefficients constituting the matrix $\boldsymbol{B}$, can be found by a least-squares procedure analogous to the one reported in Eq. (2.8). In particular, it can be demonstrated that minimizing the sum of squared residuals leads to the following values of $\boldsymbol{B}$:

$$B_{MLR} = \left(X^T X\right)^{-1} X^T Y \tag{2.17}$$

Since each of the columns of the matrix corresponds to the regression coefficients for the prediction of the corresponding column in the matrix Y, Eq. (2.17) indicates that, as far as MLR is concerned, there is absolutely no difference, when dealing with a multivariate response, in modeling all the Ys simultaneously or to consider one dependent variable at a time: in both cases, the corresponding set of coefficients would anyway be the same.

In the context of multivariate linear regression, the use of MLR has several advantages. First of all, estimation of the optimal set of model coefficients is based on a clear loss function, which represents the straightforward generalization of the least-squares criterion to the case of more than one predictor variable. Moreover, its mathematical structure is such that no additional parameters or metaparameters to be tuned in a model selection stage are involved, which translates to the fact that once the X-Y training pairs are identified, the solution is univocal. On the other hand, model calculation requires the inversion of the matrix $X^T X$, which can represent a severe limitation in many practical situations. Indeed, conditions commonly met when dealing with chemical data, such as having a lower number of training samples than of measured variables and/or a high correlation among the predictors, lead to the impossibility of inverting the matrix $X^T X$, which becomes ill-conditioned. This translates to the fact that unless some variable selection strategy is adopted (sometimes also coupled with orthogonalization approaches), MLR is seldom applied to chemical data as such.

A solution to the limitations discussed above is represented by the use of abstract (latent) variables, instead of the experimentally measured ones, as predictors. Indeed, since the chemical data are highly correlated, it is possible to exploit such correlation to achieve a substantial reduction in the dimensionality of the problem through data projection onto a low-dimensional subspace of highly informative orthogonal directions in the multivariate space (latent variables or components). The coordinates of the sample along this new set of directions (which are often referred to as *scores*) can then be used as predictors to build regression models. In particular, due to their orthogonality, scores on the different components are completely uncorrelated among one another. Moreover, the number of relevant components is usually lower than the number of training samples, allowing to overcome the limitations related to ill-conditioning of the predictor matrix.

In this framework, different methods for multivariate linear regression have been proposed in the literature, the most commonly used of which are principal component regression (PCR) (Jolliffe, 1982) and partial least-squares regression (PLS) (Wold, Martens, & Wold, 1983).

1.2 Principal component regression (PCR)

In principal component regression (Jolliffe, 1982), as the name suggests, scores along the principal components are used as predictors instead of the originally measured experimental data. Principal components (Jolliffe, 2002) are orthogonal directions in the X-space providing the best low-dimensional approximation of the original data in the least-squares sense. Their direction cosines in the variable space (also called *loadings*) are collected in the matrix $\boldsymbol{P}$, so that the corresponding scores (sample coordinates) $\boldsymbol{T}$ are given by:

$$\boldsymbol{T} = \boldsymbol{XP} \tag{2.18}$$

with $\boldsymbol{X}$ being the predictor matrix.

The scores matrix $\boldsymbol{T}$ is then used to build the MLR model instead of the original matrix $\boldsymbol{X}$:

$$\boldsymbol{Y} = \widehat{\boldsymbol{Y}} + \boldsymbol{E} = \boldsymbol{T}\,\boldsymbol{C}_{PCR} + \boldsymbol{E} \tag{2.19}$$

where the matrix of regression coefficients for the MLR model built on PCA scores was indicated as $\boldsymbol{C}_{PCR}$ to avoid confusion. Consistently with what described in Section 1.1, the optimal values of $\boldsymbol{C}_{PCR}$ are given by:

$$\boldsymbol{C}_{PCR} = \left(\boldsymbol{T}^T\boldsymbol{T}\right)^{-1}\boldsymbol{T}^T\,\boldsymbol{Y} \tag{2.20}$$

PCR is then simply a two-step modeling procedure when at first principal components are extracted from the X data and then the scores along these components are used to build a MLR model to predict the Y response(s). Anyway, since PCA scores are calculated directly from the experimental matrix through linear projection, Eq. (2.19) can be rewritten in terms of the original variables as:

$$\boldsymbol{Y} = \widehat{\boldsymbol{Y}} + \boldsymbol{E} = \boldsymbol{T}\,\boldsymbol{C}_{PCR} + \boldsymbol{E} = \boldsymbol{XP}\,\boldsymbol{C}_{PCR} + \boldsymbol{E} = \boldsymbol{X}\,\boldsymbol{B}_{PCR} + \boldsymbol{E} \tag{2.21}$$

where, analogously to what discussed in Eq. (2.16), $\boldsymbol{B}_{PCR}$ ($=\boldsymbol{P}\,\boldsymbol{C}_{PCR}$) is the matrix of regression coefficients directly relating $\boldsymbol{Y}$ to $\boldsymbol{X}$. Here, it should be stressed that despite the relation $\boldsymbol{Y}$ and $\boldsymbol{X}$ looks mathematically identical in Eqs. (2.16) and (2.21), the values of the coefficients in the two matrices $\boldsymbol{B}_{MLR}$ and $\boldsymbol{B}_{PCR}$ are numerically different. Indeed, in PCR, not all the information originally present in the $\boldsymbol{X}$ matrix is used to model the $\boldsymbol{Y}$ (and, accordingly, to estimate $\boldsymbol{B}_{PCR}$), so that the method is said to be *biased*. In particular, only the variability in the X data described by the scores $\boldsymbol{T}$ contributes to the definition of $\boldsymbol{B}_{PCR}$ and this, in turn, translates to the fact that a different model will be obtained depending on how many principal components are included in the calculation.

As a consequence, differently than what happens with MLR, building a PCR model requires the optimization of an additional metaparameter, i.e., the number of principal components to be retained. This is normally accomplished by testing models with an increasing number of principal components and selecting the one resulting in the lowest prediction error on the training data (usually computed by cross-validation or other resampling procedures, see Section 3).

PCR, as discussed above, allows the possibility of calculating regression models also in the case of ill-conditioned predictor matrices, but its two-step procedure (where each step has a different objective) makes it not necessarily the best option to rely on. Indeed, the scores extracted from the $\boldsymbol{X}$ matrix by PCA provide the best approximation of the X data, but it is not guaranteed that they are also the best predictors for the $\boldsymbol{Y}$ response(s).

On the other hand, in partial least-squares (PLS) regression, which will be described in the following paragraph, this issue is overcome since the Y data are involved already at the data compression stage to orient the latent variables, so that scores relevant for describing at the same time the variance in the predictors and the responses are extracted.

1.3 Partial least-squares regression (PLS)

In PLS (Geladi & Kowalski, 1986; Wold, Martens, & Wold, 1983; Wold, Johansson, & Cocchi, 1993; Wold, Sjöström, & Eriksson, 2001), the X and Y blocks matrices are jointly decomposed to produce scores with maximum covariance; those scores are then used as predictors for the response(s) through least-squares fitting. Accordingly, the resulting scores will represent a compromise between fitting the variance in X and approximating the dependent variable(s). On the other hand, since the Y-block is decomposed together with the predictor one, in the case of a multivariate response, building as many single-Y models as the number of dependent variables or building a single model considering all the Ys together usually leads to different results and one should decide carefully when to use which approach.

1.3.1 PLS for a single response (PLS-1)

The PLS-1 approach deals with the possibility of predicting a univariate response (y). Scores along the different components are then extracted so to have maximum covariance with the response vector and forced to be orthogonal to one another:

$$\underset{\boldsymbol{r}_a}{\operatorname{argmax}}\left\{\boldsymbol{y}^T\boldsymbol{t}_a\right\} = \underset{\boldsymbol{r}_a}{\operatorname{argmax}}\left\{\boldsymbol{y}^T\boldsymbol{X}\boldsymbol{r}_a\right\} \tag{2.22}$$

where the weight vector $\boldsymbol{r}_a$ is used to extract scores from the independent matrix ($\boldsymbol{t}_a = \boldsymbol{X}\ \boldsymbol{r}_a$) and the subscript a indicates that the condition is applied component-wise (i.e., to the ath latent variable). The X-scores ($\boldsymbol{T} = [\boldsymbol{t}_1\boldsymbol{t}_2\cdots\boldsymbol{t}_a\cdots\boldsymbol{t}_A]$, A being the number of components) are then used as predictors for the response through least-squares fitting:

$$\widehat{\boldsymbol{y}} = \boldsymbol{T}\boldsymbol{q} \tag{2.23}$$

where the regression coefficients $\boldsymbol{q}$ relating the X-scores to the Y are usually called Y-loadings, in analogy with the multivariate case. Similarly to what already highlighted for PCR, by exploiting the linear relation between the X-scores and the predictor matrix, it is possible to explicitly express the predicted response in terms of the original variable through the equation: through linear projection, Eq. (2.19) can be rewritten in terms of the original variables as:

$$\widehat{\boldsymbol{y}} = \boldsymbol{T}\boldsymbol{q} = \boldsymbol{X}\boldsymbol{R}\boldsymbol{q} = \boldsymbol{X}\ \boldsymbol{b}_{PLS} \tag{2.24}$$

where $\boldsymbol{b}_{PLS} = \boldsymbol{R}\boldsymbol{q}$, being the matrix collecting the weights for the calculation of the different components ($\boldsymbol{R} = [\boldsymbol{r}_1\ \ \boldsymbol{r}_2\ \ \cdots\ \ \boldsymbol{r}_a\ \ \cdots\ \ \boldsymbol{r}_A]$).

1.3.2 PLS for a multivariate response (PLS-2)

When also the Y data are multivariate, the PLS algorithm is modified to account for the hypothesis that both blocks share the same latent structure, i.e., that they can be described by the same low-rank approximation, which, in turn, means that the X- and Y-scores lie in the same column subspace. Mathematically speaking, these assumptions can be translated to assuming that the relevant variance in X can be summarized by the X-scores ($\boldsymbol{T} = [\boldsymbol{t}_1\ \ \boldsymbol{t}_2\ \ \cdots\ \ \boldsymbol{t}_a\ \ \cdots\ \ \boldsymbol{t}_A]$) and that the systematic variation in Y can be approximated by the Y-scores ($\boldsymbol{U} = [\boldsymbol{u}_1\ \ \boldsymbol{u}_2\ \ \cdots\ \ \boldsymbol{u}_a\ \ \cdots\ \ \boldsymbol{u}_A]$), with A being the number of extracted components:

$$\boldsymbol{T} = \boldsymbol{X}\boldsymbol{R} \tag{2.25}$$

$$U = YQ \tag{2.26}$$

where the X-weights have the same meaning as in Eq. (2.24), while $\mathbf{Q}$ $(=[\,\mathbf{q}_1\ \ \mathbf{q}_2\ \ \cdots\ \ \mathbf{q}_a\ \ \cdots\ \ \mathbf{q}_A\,])$ are the Y-loadings, which define the projection, which leads to the Y-scores $\boldsymbol{U}$. Component-wise, the X- and Y-scores are extracted so to have maximum covariance with one another:

$$\underset{\mathbf{r}_a,\mathbf{q}_a}{\operatorname{argmax}}\{\mathbf{u}_a^T\mathbf{t}_a\} = \underset{\mathbf{r}_a,\mathbf{q}_a}{\operatorname{argmax}}\{\mathbf{q}_a^T\mathbf{Y}^T\mathbf{X}\mathbf{r}_a\} \tag{2.27}$$

Here, it is important to stress that, in PLS, the linear dependence between the Y- and the X-blocks is imposed at the level of the components through the so-called inner relationship, where it is assumed that the Y-scores can be linearly approximated (component-wise) by the X-scores:

$$\widehat{\mathbf{u}}_a = \mathbf{t}_a\, c_{a,PLS} \tag{2.28}$$

with $c_{a,\ PLS}$ being the proportionality constant (which is estimated by univariate least squares) and the hat, indicating that the Y-scores are predicted from the X ones. In the matrix form, Eq. (2.28) can be rewritten as:

$$\widehat{\boldsymbol{U}} = \boldsymbol{T}\,\boldsymbol{C}_{PLS} \tag{2.29}$$

with $\boldsymbol{C}_{PLS}$ being a diagonal matrix, whose elements are the various $c_{a,PLS}$.

Predicted responses are then obtained from the estimated Y-scores by inverting the relation expressed by Eq. (2.26), i.e., through the Y-loadings:

$$\widehat{\mathbf{Y}} = \widehat{\boldsymbol{U}}\,\mathbf{Q}^T \tag{2.30}$$

Also in the case of a multivariate Y, it is possible to express the overall regression model in terms of the original X variables, by properly combining the terms in Eqs. (2.25), (2.29), and (2.30):

$$\widehat{\mathbf{Y}} = \widehat{\boldsymbol{U}}\,\mathbf{Q}^T = \boldsymbol{T}\,\boldsymbol{C}_{PLS}\,\mathbf{Q}^T = \boldsymbol{X}\boldsymbol{R}\,\boldsymbol{C}_{PLS}\,\mathbf{Q}^T = \boldsymbol{X}\,\boldsymbol{B}_{PLS}, \tag{2.31}$$

where the regression matrix relating the Y- and the X-block is given by:

$$\boldsymbol{B}_{PLS} = \boldsymbol{R}\,\boldsymbol{C}_{PLS}\,\mathbf{Q}^T. \tag{2.32}$$

Since not all the information in the X-block is used for predicting the Y, PLS, as already discussed in the case of PCR, represents another example of the so-called biased regression, which, for model building, requires to define how many components should be used for the prediction. Analogously to what already discussed for PCR, the selection of the optimal number of components is usually accomplished by choosing the value leading to the minimum prediction error (usually, in cross-validation).

1.4 Other regression techniques

Although for many chemical problems the use of a linear regression model is often sufficient to provide reliable predictions and, in these cases, the large majority of the applications involve the use of one of the three techniques discussed above, sometimes the functional relation to be approximated may be more complex and non-linear methods should be adopted. Since artificial neural networks (ANN) (Marini, Bucci, Magrì, & Magrì, 2008) and support vector machines (SVM) (Brereton & Lloyd, 2010) are described in other chapters of this book, here we will just briefly mention how to implement non-linearities in the context of the models described so far.

In the case of PCR, a non-linear relation between the Y- and the X-block can be accomplished by augmenting the scores matrix with quadratic and interaction terms, prior to using it for estimating the MLR model. For instance, if three PCs are extracted from the X-block, the scores matrix can be augmented as follows:

$$\boldsymbol{T}_{aug} = \begin{bmatrix} \boldsymbol{t}_1 & \boldsymbol{t}_2 & \boldsymbol{t}_3 & \boldsymbol{t}_1^2 & \boldsymbol{t}_2^2 & \boldsymbol{t}_3^2 & \boldsymbol{t}_1 \circ \boldsymbol{t}_2 & \boldsymbol{t}_1 \circ \boldsymbol{t}_3 & \boldsymbol{t}_2 \circ \boldsymbol{t}_3 \end{bmatrix} \tag{2.33}$$

where the symbol $\circ$ indicates Hadamard (element-wise) product of the two vectors. Then, the predicted response is estimated as a linear combination of the columns of $\boldsymbol{T}_{aug}$.

On the other hand, in the case of PLS (Berglund & Wold, 1997), non-linearity can be imposed at the level of the inner relation where, instead of Eq. (2.28), any non-linear function f can be used to approximate the Y-scores from the X-scores (Wold, Kettaneh-Wold, & Skagerberg, 1989):

$$\widehat{\boldsymbol{u}}_a = f(\boldsymbol{t}_a). \tag{2.34}$$

Moreover, the same kernel trick used in the context of support vector machines (SVM) can also be coupled to PLS regression, leading to kernel PLS (Rosipal, 2003).

2. Classification

In chemometrics, "classifying" refers to the possibility of creating models that allow assigning samples to *classes* (or *categories*) on the basis of the chemical data collected for this purpose. The aims of classification are manifold; just to mention some, in the context of quality control, classifiers can be applied to establish whether samples deviate from compliant batches or not; otherwise, in forensic science, classification approaches are widely used in order to prevent frauds (e.g., food or drug adulteration).

Classifiers belong to the *pattern recognition* methodologies, which can either be *unsupervised,* when they do not need any a priori information about class-membership of samples, or *supervised,* when a training set of objects, whose actual class-belonging is known, is needed. In particular, classification techniques fall within the latter category of methods, while for more information on unsupervised approaches the reader is addressed to (Albalate & Minker, 2013).

Classification methods can be divided into two main families: *discriminant* and *class-modeling* approaches (Cocchi, Biancolillo, & Marini, 2018). The main difference between these two categories lays in the way the hyperspace of samples is divided into regions associated with the various classes. Applying discriminant approaches, the entire sample space is divided in as many regions as the number of classes present in the investigated data set. Fig. 2.1 depicts how the sample hyperspace is divided when a three-class problem is solved by a discriminant (left plot) or a class-modeling (right plot) approach. In the former case, training samples belong to class A, class B, and class C, and the sample space is divided into three regions. As a consequence, each object will be uniquely assigned to the class-region it falls into. On the other hand, using class-modeling methods (right plot), each category is individually modeled, with the consequence that class-regions could overlap and/or not cover all the samples' space. A direct consequence of this is that some individuals will fall in one class-region and they will be assigned to that class, some other could fall in an overlapping region, and they will be accepted by more than one class-model (*confused samples*), and, eventually, some individuals could be rejected by all the class-models (i.e., they will fall outside any class-region) and being labeled as outliers for all the categories.

A brief description of some of the most common classifiers is provided in the subparagraphs below.

2.1 Linear discriminant analysis (LDA)

One of the first, but still widely used, discriminant classifiers is linear discriminant analysis proposed by Fisher (1936). LDA is a discriminant parametric approach seeking the linear hyperplanes which allow dividing the samples' space in an appropriate number of regions. As a parametric approach, it lays on the Bayes' rule, and it is based on two main postulations:

1. *Each investigated individual should be assigned to the class it is most probable it belongs to*
2. *Within each category, observations are normally distributed*

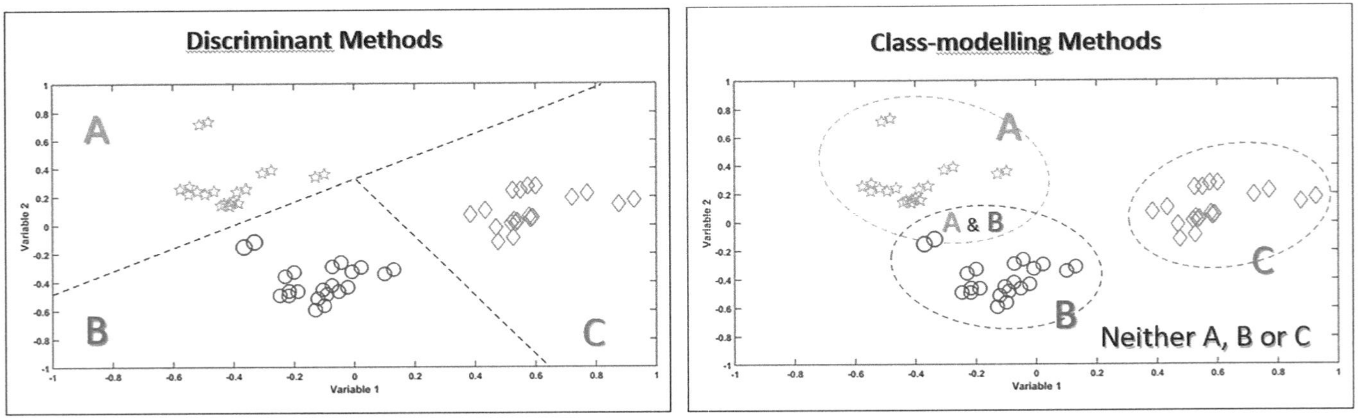

Fig. 2.1 Graphical illustration of the differences between discriminant and modeling approaches to classification.

The assumption (1) derives from Bayes' rule and is in common among parametric classifiers. It means that, considering a sample vector $\boldsymbol{x}_i$, the probability an *i*th object belongs to a class *g* is given by the *posterior probability* ($p(g\,|\,\boldsymbol{x}_i)$), which is proportional to the *likelihood* ($p(\boldsymbol{x}_i\,|\,g)$) and the a priori probability $p_0(g)$. What discriminates parametric classifiers from one and another is the way likelihood is estimated. In LDA, this is done in agreement with Eq. (2.35):

$$p(x_i|g) = \frac{1}{(2\pi)^{M/2}|S|} e^{-0.5\left(x_i-\overline{x}_g\right)^T S^{-1}\left(x_i-\overline{x}_g\right)} \tag{2.35}$$

where M is the number of measured variables, $\mathbf{S}$ is the *pooled variance/covariance matrix* (calculated as $\frac{\sum_{g=1}^{G}(n_g-1)\mathbf{S}_g}{N-G}$ where $\mathbf{S}_g$ is the variance/covariance matrix of the *g*th class and G is the total number of classes, while n_g and N are the number of samples in the *g*th category and the total number of individuals, respectively), and $\overline{\mathbf{x}}_g$ is the centroid of the considered category.

For the easiest classification problem possible, i.e., when only two categories (A and B) are considered, the sample space is divided into two class-regions by the individual boundary which verifies Eq. (2.36):

$$p\ (A|\boldsymbol{x}_i) = p\ (B|\boldsymbol{x}_i) \tag{2.36}$$

Meaning that a borderline sample has the same posterior probability of belonging to either class A or class B. It follows that the objects satisfying $p\ (\boldsymbol{x}_i|\,A) > p\ (\boldsymbol{x}_i|\,B)$ will be assigned to class A, while those for which $p\ (\boldsymbol{x}_i|\,B) > p(\boldsymbol{x}_i|\,A)$, to class B.

For each category, it is possible to estimate *classification probability functions* ($\phi_g(\boldsymbol{x})$). In fact, taking the logarithm of Eq. (2.35), $\phi_A(\boldsymbol{x})$ and $\phi_B(\boldsymbol{x})$ will correspond to:

$$\begin{aligned}\phi(\boldsymbol{x}) &= \ln p(A|\boldsymbol{x}) \\ &= \ln \frac{\Gamma}{(2\pi)^{M/2}|\mathbf{S}|} + \ln p_o(A) - 0.5\,\mathbf{x}^T\,\mathbf{S}^{-1}\,\mathbf{x} + \overline{\mathbf{x}}_A^T\,\mathbf{S}^{-1}\,\mathbf{x} \\ &-0.5\,\overline{\mathbf{x}}_A^T\,\mathbf{S}^{-1}\,\overline{\mathbf{x}}_A\end{aligned} \tag{2.37}$$

$$\begin{aligned}\phi_B\ (\boldsymbol{x}) &= \ln p(B|\boldsymbol{x}) \\ &= \ln \frac{\Gamma}{(2\pi)^{M/2}|\mathbf{S}|} + \ln p_o(B) - 0.5\,\mathbf{x}^T\,\mathbf{S}^{-1}\,\mathbf{x} + \overline{\mathbf{x}}_B^T\,\mathbf{S}^{-1}\,\mathbf{x} \\ &-0.5\,\overline{\mathbf{x}}_B^T\,\mathbf{S}^{-1}\,\overline{\mathbf{x}}_B\end{aligned} \tag{2.38}$$

where Γ is a normalization factor, ensuring that the probabilities sum up to 1.

By equating the two expressions and rearranging, it can be shown that the boundary can be estimated as:

$$\ln \frac{\Gamma p_0(A)}{\Gamma p_0(B)} - 0.5(\overline{x}_A - \overline{x}_B)^T \mathbf{S}^{-1}(\overline{x}_A - \overline{x}_B) + (\overline{x}_A - \overline{x}_B)^T \mathbf{S}^{-1}\mathbf{x}{=}0 \quad (2.39)$$

Finally, including all the constant terms in w_0 and the coefficients of the linear combinations in $\mathbf{w}$, the equation defining the class-boundary can be expressed as:

$$w_0 + \mathbf{w}^T \mathbf{x} = 0 \quad (2.40)$$

This latter equation unveils that LDA is a projection method seeking for the directions which maximize the distance between the class-centroids (*inter-class variability*) and minimize the within-variability in each category (*intra-class variability*). In the LDA context, these directions (i.e., *latent variables*) are also known as *canonical variates* (Krzanowski, 2000).

The same reasoning can also be extended for the solution of multi-class problems; for $G > 2$, a sample is predicted as belonging to one category only if its posterior probability of belonging to that class is higher than the probability of belonging to any other:

$$p(H|\mathbf{x}) > p(I|\mathbf{x}) \quad \forall I = 1 \ldots G, H \neq I \quad (2.41)$$

Boundaries can be defined similarly as before, but by pairs of classes:

$$\ln \frac{\Gamma p_0(H)}{\Gamma p_0(I)} - 0.5(\overline{\mathbf{x}}_{\mathrm{H}} - \overline{\mathbf{x}}_{\mathrm{I}})^T \mathbf{S}^{-1}(\overline{\mathbf{x}}_{\mathrm{H}} - \overline{\mathbf{x}}_{\mathrm{I}}) + (\overline{\mathbf{x}}_H - \overline{\mathbf{x}}_{\mathrm{I}})^T \mathbf{S}^{-1}\mathbf{x} = 0$$
$$\forall I = 1 \ldots G, H \neq I \quad (2.42)$$

As previously stated, canonical variates represent the direction which allows maximizing the ratio between the inter- and the intra-class variability.

In the multi-class, case is necessary to define a between-group variance/covariance matrix $\boldsymbol{S}_{\mathrm{b}}$:

$$\boldsymbol{S}_{\mathrm{b}} = \frac{1}{G} \sum_{g=1}^{G} (\overline{\mathbf{x}}_{\mathrm{G}} - \overline{\mathbf{x}})(\overline{\mathbf{x}}_G - \overline{\mathbf{x}})^T \quad (2.43)$$

Once $\boldsymbol{S}_{\mathrm{b}}$ has been defined for all categories, the solution of the eigenvalue problem:

$$\mathbf{S}^{-1} \boldsymbol{S}_{\mathrm{b}}\, \boldsymbol{w}_{\mathrm{i}} = \lambda_i\, \boldsymbol{w}_{\mathrm{i}} \quad (2.44)$$

allows the estimation of the canonical variates ($\boldsymbol{w}_{\mathrm{i}}$).

Eq. (2.44) clearly shows the main disadvantage of LDA, i.e., the fact $\boldsymbol{S}$ has to be an invertible matrix. This condition is rarely met, in particular working with instrumental data, because, among the other constrains, it requires the number of the objects to be higher than the number of the variables; nevertheless, efficient alternatives have been proposed. Among the discriminant approaches proposed to overcome this issue, partial least-squares discriminant analysis (PLS-DA) is probably one of the most commonly used and well-known approaches.

2.2 Partial least-squares discriminant analysis (PLS-DA)

One of the possible solutions conceived to overcome the invertibility issues associated with LDA is to include a feature reduction step prior to the application of this classifier. This can be achieved by exploiting a variable selection method (for instance, a stepwise procedure) before calculating the LDA model. Another alternative is the application of methods based on latent variables (for example, PCA), which compress the information into a restricted subset of orthogonal features and make the calculation of LDA decision boundaries possible.

Eventually, PLS-DA exploits a further strategy; the classification problem is transformed into a regression one, which can be therefore solved by PLS (Barker & Rayens, 2003; Sjöström, Wold, & Söderström, 1986; Ståhle & Wold, 1987). This transition is made possible by the introduction of a *dummy* $\boldsymbol{Y}$, i.e., a $\boldsymbol{Y}$-response which binary codes the class-membership of training samples. The codification is easy: each sample is represented by a row-vectors made of zeros except for the column corresponding to its class-membership. For instance, for a three-class case, samples belonging to class A, class B, and class C can be represented by the row-vectors $\boldsymbol{y}_A^T = [1\ 0\ 0]$; $\boldsymbol{y}_B^T = [0\ 1\ 0]$; $\boldsymbol{y}_c^T = [0\ 0\ 1]$, respectively.

Once $\boldsymbol{Y}$ is created, the general regression equation (Eq. 2.6) can be solved by PLS and the calibration model built. This step includes the estimation of the regression coefficient matrix $\boldsymbol{B}$, which allows the prediction of unknown samples. In this case, new observations can be collected in a novel data matrix $\boldsymbol{X}_{new}$ and the response matrix $\widehat{\boldsymbol{Y}}_{new}$ estimated (as already described in Section 1.3) by:

$$\widehat{\boldsymbol{Y}}_{new} = \boldsymbol{X}_{new}\boldsymbol{B} \tag{2.45}$$

Contrarily to the target responses $\boldsymbol{Y}$, the predicted ones $\widehat{\boldsymbol{Y}}_{new}$ are not binary, but made of continuous variables, and a classification rule has to be defined in order to predict the class-membership of new samples. On this regard, different solutions have been proposed (Indahl, Martens, & Næs, 2007; Pérez, Ferré, & Boqué, 2009). Here, it should be pointed out that, for the two-class case, due to the symmetry of the discriminant classification problem (if a sample is not class A, then it should necessarily be class B), the dummy matrix $\boldsymbol{Y}$ becomes a dummy vector $\boldsymbol{y}$: the target response becomes either 1, if the sample belongs to the first class, or 0 if it belongs to the second one, Then, classification is accomplished by setting a threshold $\widehat{y}_{thr}$ to the values of the predicted response $\widehat{y}_{new}$ so that if $\widehat{y}_{new} > \widehat{y}_{thr}$ the sample is predicted as class 1; otherwise if $\widehat{y}_{new} < \widehat{y}_{thr,}$ it is predicted as class 2. Accordingly, a value of $\widehat{y}_{thr} = 0.5$ if often naively used, but more efficient approaches for the estimation of the best value of $\widehat{y}_{thr}$ are discussed in the literature. When more than two classes are involved, the most natural and straightforward approach for implementing classification based on the predicted responses $\widehat{\boldsymbol{Y}}_{new}$ is as follows. Each new sample is predicted to belong to the class corresponding to the component of the predicted response with the highest value. For instance, taking into consideration the matrix:

$$\widehat{\boldsymbol{Y}}_{New} = \begin{bmatrix} \widehat{\boldsymbol{y}}_1^T \\ \widehat{\boldsymbol{y}}_2^T \\ \widehat{\boldsymbol{y}}_3^T \end{bmatrix} = \begin{bmatrix} +0.4 & -0.2 & 1.2 \\ 0.3 & 0.9 & 0.4 \\ 1.1 & 0.2 & -0.3 \end{bmatrix} \tag{2.46}$$

The first sample, based on the predicted response $\widehat{\boldsymbol{y}}_1^T$,will be predicted as belonging to class C, because the element of $\widehat{\boldsymbol{y}}_1^T$ with the highest value (1.2) is in the third column; the second individual will be assigned to class B, because the highest value of its predicted response $\widehat{\boldsymbol{y}}_2^T$ (0.9) is in the second column; lastly, the third sample will be predicted as belonging to class A, because the highest value of its predicted response $\widehat{\boldsymbol{y}}_3^T$ (1.1) is in the first column. As anticipated above, other approaches have been proposed in the literature to build classification decision rules based on the predicted response in PLS-DA. One of the most commonly used approaches is to apply LDA directly on the predicted responses or on the PLS scores (Indahl et al., 2007).

2.3 Soft independent modeling of class analogies (SIMCA)

Although the discriminant methods previously mentioned perform very well from a predictive point of view, in some cases, they are not always the most appropriate choice for all the situations that may occur. For example, in the case of *asymmetric classification*, i.e., when the focus is not on assigning all samples to one of the possible classes, but on modeling only a specific category of interest, class-modeling methods may be more appropriate than the discriminant ones. In fact, these methods allow the individual modeling of the category of interest, and then all the investigated individuals can be accepted (and, therefore, predicted as belonging to the class of interest) or rejected (i.e., they are predicted as not belonging to the modeled class without any further cogitation on its possible class-membership) by the model. Soft independent modeling of class analogies (SIMCA) is probably the most well-known class-modeling approach; it has been proposed by Wold in 1977 (Wold, 1976; Wold & Sjöström, 1977), and, since then, various modifications have been proposed to its original formulation, in order to improve its efficiency (Derde & Massart, 1988; Pomerantsev & Rodionova, 2020).

Basically, SIMCA starts from the assumption that the main characteristics of the samples belonging to the class of interest can be captured by a principal component analysis (PCA) model of proper dimensionality. Consequently, the first step of the algorithm consists of calculating a PCA model on the training individuals belonging to the class of interest:

$$\boldsymbol{X}_g = \boldsymbol{T}_g \boldsymbol{P}_g^T + \boldsymbol{E}_g \tag{2.47}$$

where the symbols have the same meaning as in Eq. (2.18), while the subscript g indicates that the model is calculated only on the training samples of class g. $\boldsymbol{E}_g$ indicates the residuals after having approximated the matrix $\boldsymbol{X}_g$ with A_g principal components. Here, it should be stressed that the number of principal components is a meta-parameter which should be optimized. This is usually accomplished by testing models with an increasing number of components and selecting the dimensionality, which leads to the highest classification efficiency (usually in cross-validation, see Section 3).

Based on the projection expressed by Eq. (2.47), a distance of any individual to the model is calculated as a combination of its orthogonal distance (i.e., the residuals after PCA projection, indicated as Q_i) and its (usually Mahalanobis) distance to the center of the scores space (indicated as T_i^2). The various implementations of SIMCA proposed in the literature differ in the way they combine scores and orthogonal distances into a single distance to the model and, as it will be discussed below, in the way they set a threshold to the overall distance in order to define acceptance or rejection by the class model (Pomerantsev & Rodionova, 2020).

The most popular formulation of SIMCA defines the overall distance to the model as:

$$d_i = \sqrt{\left(\frac{T_i^2}{T_{0.95}^2}\right)^2 + \left(\frac{Q_i}{Q_{0.95}}\right)^2} \tag{2.48}$$

where the subscript 0.95 indicates that both the scores and the orthogonal distances, prior to being combined, need to be normalized by division to the 95th percentiles of their respective distributions under the null hypothesis.

When the overall distance to the model is defined as in Eq. (2.48), as a consequence of the normalization involved, its threshold for acceptance is set to $\sqrt{2}$. Therefore, if the i-th sample presents a $d_i \leq \sqrt{2}$, the object is accepted by the model and predicted as belonging to the modeled category; otherwise, it is rejected and considered not belonging to the class of interest.

3. Validation

Chemometrics modeling is of empirical nature, i.e., models are built based on the collected data, so to provide approximations of the system under study and formulate reliable predictions. In this context, it is necessary to point out that, given the available data, the possible models that can be calculated are almost always more than a single one, and the quality and the performances of the concurrent candidate models can be influenced by several factors, e.g., the number of samples and their representativeness, the characteristics of the method or the algorithm for calculating the solution. In this context, it is fundamental to assess whether reliable outcomes and predictions can be drawn for a particular model, and whether they can be generalizable or not, at the same time verifying whether the interpretation of the results could be sound and meaningful. This process is called validation, and it constitutes a fundamental step of any chemometric modeling (Westad & Marini, 2015).

Operationally, validation is often conducted by applying the model to a set of data (test set) which should be different and independent from the one used for model building (what was defined, in the text as the training or calibration set). This additional set of data is treated along all the analyses as if it were a new set of unknown observations, so to be able to obtain the model predictions for those data, but it is also necessary that the true values of the quantities to be predicted are known as well, in order to evaluate the model performances and to be able to calculate various possible diagnostics based on the value of the residuals (i.e., the differences between the true and predicted values of the responses). In order for the validation to be effective, the use of new data instead of the training samples is of the highest importance. Indeed, since the model parameters are often optimized so to produce the minimum error on the training observation, by using the same data for model validation, one would obtain over-optimistic results by definition. On the other hand, the test set should be chosen so to reflect as closely as possible all the possible sources of variability that one would expect in the routine application of the model, i.e., on future individuals. In this context, when collecting a completely new set of samples is not possible, a solution is represented by the possibility of splitting all the available data into training and test sets, possibly not randomly but through the use of algorithms which guarantee that the two sets span as effectively as possible the sample space [e.g., Kennard-Stone (Kennard & Stone, 1969) or Duplex (Snee, 1977)].

When the number of samples is limited, resampling strategies, like cross-validation (CV), can also be employed. In cross-validation, the whole data set is divided into a specified number of cancelation groups, which, in turn, are used as validation sets, while the remaining individuals constitute the training set. CV has the advantage that any sample is predicted by a model which was built on a set that doesn't include it; on the other hand, since training and test sets are never completely independent from one another, the results that are obtained could in general be too optimistic. Anyway, cross-validation, when applied to the training samples only, can be fruitfully exploited in the model selection stage, for instance, when the optimal values of some metaparameters, e.g., the number of components, should be identified. In that case, one can compare the performances, in CV, of the various competing model and then select the value of the metaparameter(s) leading to the best performances. Once the optimal model has been selected, validation of its performances on a completely external test set is anyway recommended.

References

Albalate, A., & Minker, W. (2013). *Semi-supervised and unsupervised machine learning: Novel strategies*. John Wiley and Sons. https://doi.org/10.1002/9781118557693.

Barker, M., & Rayens, W. (2003). *Journal of Chemometrics*, *17*(3), 166–173. https://doi.org/10.1002/cem.785.

Berglund, A., & Wold, S. (1997). *Journal of Chemometrics*, *11*(2), 141–156. https://doi.org/10.1002/(SICI)1099-128X(199703)11:2<141::AID-CEM461>3.0.CO;2-2.

Brereton, R. G., & Lloyd, G. R. (2010). *Analyst, 135*(2), 230–267. https://doi.org/10.1039/b918972f.
Cocchi, M., Biancolillo, A., & Marini, F. (2018). *Comprehensive analytical chemistry. Vol. 82* (pp. 265–299). Elsevier B.V. https://doi.org/10.1016/bs.coac.2018.08.006.
Derde, M. P., & Massart, D. L. (1988). *Chemometrics and Intelligent Laboratory Systems, 4*(1), 65–93. https://doi.org/10.1016/0169-7439(88)80013-3.
Fisher, R. A. (1936). *Annals of Eugenics*, 7(2), 179–188. https://doi.org/10.1111/j.1469-1809.1936.tb02137.x.
Geladi, P., & Kowalski, B. R. (1986). *Analytica Chimica Acta, 185*(C), 1–17. https://doi.org/10.1016/0003-2670(86)80028-9.
Indahl, U. G., Martens, H., & Næs, T. (2007). *Journal of Chemometrics, 21*(12), 529–536. https://doi.org/10.1002/cem.1061.
Jolliffe, I. T. (1982). *Applied Statistics, 31*(3), 300. https://doi.org/10.2307/2348005.
Jolliffe, I. T. (2002). *Principal component analysis*. Springer. https://link.springer.com/book/10.1007/b98835.
Kennard, R. W., & Stone, L. A. (1969). *Technometrics, 11*(1), 137–148. https://doi.org/10.1080/00401706.1969.10490666.
Krzanowski, W. J. (2000). *Principles of multivariate analysis: A user's perspective* (2nd ed.). Oxford University Press. https://global.oup.com/academic/product/principles-of-multivariate-analysis-9780198507086?cc=it&lang=en&.
Marini, F., Bucci, R., Magrì, A. L., & Magrì, A. D. (2008). *Microchemical Journal, 88*(2), 178–185. https://doi.org/10.1016/j.microc.2007.11.008.
Martens, H., & Naes, T. (1991). *Multivariate calibration*. John Wiley & Sons.
Pérez, N. F., Ferré, J., & Boqué, R. (2009). *Chemometrics and Intelligent Laboratory Systems, 95*(2), 122–128. https://doi.org/10.1016/j.chemolab.2008.09.005.
Pomerantsev, A. L., & Rodionova, O. Y. (2020). *Journal of Chemometrics, 34*(8). https://doi.org/10.1002/cem.3250.
Rosipal, R. (2003). *Neural Network World, 13*(3), 291–300.
Sjöström, M., Wold, S., & Söderström, B. (1986). PLS discriminant plots. In E. S. Gelsema, & L. N. Kanal (Eds.), *Pattern recognition in practice* (pp. 461–470). Amsterdam: Elsevier.
Snee, R. D. (1977). *Technometrics, 19*(4), 415–428. https://doi.org/10.1080/00401706.1977.10489581.
Ståhle, L., & Wold, S. (1987). *Journal of Chemometrics, 1*(3), 185–196. https://doi.org/10.1002/cem.1180010306.
Westad, F., & Marini, F. (2015). *Analytica Chimica Acta, 893*, 14–24. https://doi.org/10.1016/j.aca.2015.06.056.
Wold, S. (1976). *Pattern Recognition, 8*(3), 127–139. https://doi.org/10.1016/0031-3203(76)90014-5.
Wold, S., Johansson, E., & Cocchi, M. (1993). PLS: Partial least squares projections to latent structures. In H. Kubinyi (Ed.), *3D QSAR in drug design: Theory, methods and applications* (pp. 523–550). Leiden, The Netherlands: ESCOM Science Publisher.
Wold, S., Kettaneh-Wold, N., & Skagerberg, B. (1989). *Chemometrics and Intelligent Laboratory Systems*, 7(1–2), 53–65. https://doi.org/10.1016/0169-7439(89)80111-X.
Wold, S., Martens, H., & Wold, H. (1983). The multivariate calibration problem in chemistry solved by the PLS method. In A. Ruhe, & B. Kågström (Eds.), *Proceedings of the conference on matrix pencils. Lecture notes in mathematics* (pp. 286–293). Heidelberg, Germany: Springer Verlag.
Wold, S., & Sjöström, M. (1977). SIMCA: A method for analyzing chemical data in terms of similarity and analogy. In B. R. Kowalski (Ed.), *Chemometrics: Theory and application* (pp. 243–282). Washington, DC: American Chemical Society.
Wold, S., Sjöström, M., & Eriksson, L. (2001). *Chemometrics and intelligent laboratory systems. Vol. 58* (pp. 109–130). https://doi.org/10.1016/S0169-7439(01)00155-1. Issue 2.

CHAPTER 3

Multivariate pattern recognition by machine learning methods

Parvin Razzaghi[a], Karim Abbasi[b], and Jahan B. Ghasemi[c]
[a]Department of Computer Science and Information Technology, Institute for Advanced Studies in Basic Sciences (IASBS), Zanjan, Iran
[b]Laboratory of System Biology, Bioinformatics & Artificial Intelligent in Medicine (LBB&AI), Faculty of Mathematics and Computer Science, Kharazmi University, Tehran, Iran
[c]Faculty of Chemistry, University of Tehran, Tehran, Iran

1. Introduction

In pattern recognition, the goal is to identify patterns and regularities from data with minimal intervention (Bishop, 2006). Finding the patterns and regularities is done using machine learning algorithms. In other words, pattern recognition is the engineering application, while machine learning origin back in computer science. Machine learning algorithms map the input data to the output, and they consist of two main steps: the training step and the inference step. In the training step, the training data is utilized to learn a model, and in the inference step, the learned model is used to map unseen data to the output space. Machine learning algorithms are divided into four categories from the view of the available training data: (1) supervised learning, (2) semi-supervised learning, (3) weakly supervised learning, and (4) unsupervised learning. In supervised learning, all the available training data are fully annotated (labeled) as compared to unsupervised learning in which none of the training data is labeled. In semi-supervised learning, some portion of training data is labeled and a large portion is unlabeled. If the whole training data is not fully labeled (i.e., partially labeled), then it is weakly supervised learning. In this chapter, we focus on supervised learning and in each section.

Given $\{(x^{(i)}, y^{(i)})\}_{i=1}^{N}$ denotes the training samples where $x^{(i)} \in \chi$ and $y^{(i)} \in \Upsilon$ show the input and output of the ith sample, respectively. The main goal is to learn a model which maps an unseen data x^{test} from the input space to the output space. The overall schematic of the machine learning approaches is shown in Fig. 3.1. As it is represented, the first step of machine learning algorithms is feature extraction in which important and discriminative knowledge from the raw data is extracted. After extracting features for

Machine Learning and Pattern Recognition Methods in Chemistry from Multivariate and Data Driven Modeling
https://doi.org/10.1016/B978-0-323-90408-7.00002-2

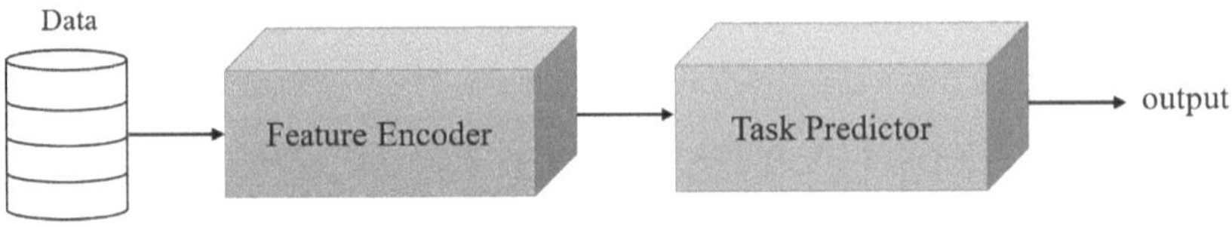

Fig. 3.1 The overall schematic of the proposed approach.

each input data, the output prediction is done. In the following, the problem formulation of the machine learning algorithms is given.

The feature extraction step takes the raw data as input and maps it to feature vector, which is given as follows:

$$\varphi = \phi(x) \tag{3.1}$$

where ϕ is the feature extractor function. In the prediction task, the obtained feature vector is obtained as input and maps it to the output space, which is denoted as follows:

$$\hat{y} = h(\phi(x); \theta) \tag{3.2}$$

where h is a predictor (hypothesize) function and θ is the set of parameters that model the predictor function. Now, there is a set of unknown parameters which should be obtained during the learning stage. To obtain the parameters, the loss function or cost function should be defined. Loss function represents the cost associated with the specific value of the parameters, which is defined as:

$$L(\theta) = \sum_i l(y_i, \widehat{y}_i) \tag{3.3}$$

where l (.,.) computes how the predicted output $(\widehat{y})$ is close or far from the real output value (y). By optimizing the loss function, the optimal value for the parameter is obtained, which is given as follows:

$$\theta^* = \underset{\theta}{argmin}\, L(\theta) \tag{3.4}$$

In this chapter, the machine learning algorithms are divided into two categories based on how the feature extraction step is done: classic approaches and deep learning-based approaches. In the following, at first, feature extraction ad prediction task steps are explained in detail, and then, classic approaches and deep learning-based approaches are given.

2. Feature extraction

The first step of machine learning algorithms is feature description. In this step, the interesting information of input is encoded into a series of numbers, which is commonly called a fingerprint. This step could be done in two different ways: (1) non-data-driven-based feature extraction and (2) data-driven-based feature extraction. In a non-data-driven-based way, the interesting information of input is extracted according to a certain manually predefined algorithm by utilizing the expert knowledge. These features are also called hand-crafted features or hand-engineered features. In the following, as an example, we take the example of molecular property prediction to better explain concepts and algorithms. In molecular property prediction, the input is a small molecule compound which some of these molecules are shown in Fig. 3.2. The goal is to build machine learning algorithms that take these small-molecule compounds as input and map them to the drug's activity property. In drug molecule property prediction, small-molecule compounds (as input) could be described using a non-data-driven-based algorithm. In the literature, there are many non-data-driven-based small-molecule description algorithms, which include extended connectivity fingerprint (ECFP) (Rogers & Hahn, 2010), ToxPrint (Yang et al., 2015), and molecular access system (MACCS) (Durant, Leland, Henry, & Nourse, 2002). ECFP is a topological molecule descriptor in which features are extracted by means of circular atom neighborhoods. MACCS is a frequently used m-bit 2D structure fingerprint where each bit encodes the presence of a specific substructure in the molecule (Durant et al., 2002). The ToxPrint utilizes a library of more than 700 chemotypes in which each molecule is represented by substructures associated with toxic effects, and it is commonly important for toxicity predictions.

As compared to non-data-driven-based descriptors, in data-driven-based algorithms, the representation for each molecule is learned automatically. In recent years, a large number of data-driven-based algorithms have been developed (David, Thakkar, Mercado, & Engkvist, 2020). Abbasi, Razzaghi, Poso, Ghanbari-Ara, and Masoudi-Nejad (2021) state that six data-driven feature extraction architectures get the most attention in the

Acetaminophen | Methacetin methoxy-C-13 | Diloxanide

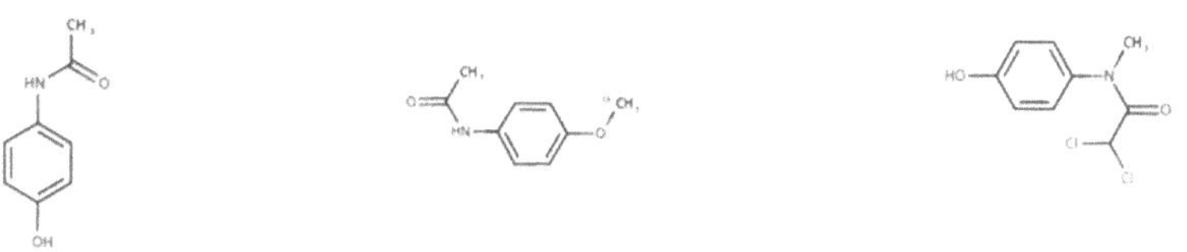

CC(=O)NC1=CC=C(O)C=C1 | CC(=O)NC1=CC=C(O[13CH3])C=C1 | CN(C(=O)C(Cl)Cl)C1=CC=C(O)C=C1

Fig. 3.2 Some examples of compound molecules. In this figure, the name of the molecule, its 2D structure, and SMILES sequence are given.

literature, namely, (1) convolutional neural network (CNN), (2) graph convolutional network (GCN), (3) recurrent neural network (RNN), (4) autoencoder, (5) deep belief network (DBN), and (6) word-embedding methods. All of these algorithms try to learn the important substructure from the input sequence, which is important in associated task prediction.

3. Task prediction

In this section, task prediction is explained. The obtained feature descriptor of the input from the previous step is fed as input into the task prediction. In a machine learning problem, a task is a type of prediction or inference which is being asked in the problem. Each problem could have many types of tasks. Based on the literature, we have divided different tasks into four categories: (1) regression, (2) classification, (3) ranking, and (4) structural output prediction. Regression assigns a continuous label to each sample. For example, in compound-target prediction, a real binding affinity value is assigned to a pair of compound and protein targets, which states the strength of the binding interaction of a biomolecule target (e.g., protein, enzyme, or DNA) with the ligand of the compound. In classification, a discrete label is assigned to each input sample, which states the class label of the input data. The classification task is divided into two categories: binary classification and multiclass classification. In binary classification, the number of classes is two, and in multiclass classification, the number of classes is more than two. Prediction of drug-likeness is an example of binary classification in which the input drug compound is divided into drug-like/non-drug-like class labels. Ranking assigns a partial order for each item by giving a numerical or ordinal score, which states the relevance degree of each item match. Gerdes et al. (2021) utilize omics data to assign a numerical score to lists of drugs based on their antiproliferative efficacy in the cancer cell. In the structural output prediction task, a structured object is predicted as output compared to classification and regression, which produce, respectively, scalar discrete or real values as an output. For example, protein 3D structure prediction (Jumper et al., 2021) takes the protein sequence as input, then produces the 3D structure of the protein as output. In this chapter, we focus on regression and classification tasks.

3.1 Classic approaches

In this section, the classic approaches for regression and classification tasks are investigated. In classic machine learning algorithms, to extract features from the raw input data, the hand-crafted feature extractor algorithms are utilized, and then, it is fed into the task predictor.

In this section, to explain the introduced algorithms in a practical way, we take the molecular property prediction problem as an application. In molecule property prediction, the input is a compound and the output could be the measured log solubility in mols per liter (for regression) or activity/non-activity (for classification). The drug could be represented as a SMILES sequence or 2D graph structure. Also, in this section, the ECFP algorithm is used to describe the input drug.

3.2 Regression

In this section, the linear regression algorithm is investigated. In regression, the output space Υ is continuous. In linear regression, the predictor function (i.e., h) is defined as follows:

$$h(\phi(x);\theta) = \sum_i \varphi_i \theta_i \tag{3.5}$$

As it is shown, the predictor function is a sum of the linear weighted of the obtained features from function ϕ (x). Also, in linear regression algorithm, the loss function is defined as follows:

$$\begin{aligned} L\left(\theta; \left\{\left(x^{(i)}, y^{(i)}\right)\right\}_{i=1}^{N}\right) &= \sum_i \left\| y^{(i)} - h\left(\phi\left(x^{(i)}\right);\theta\right)\right\|^2 \\ &= \sum_i \left\| y^{(i)} - \hat{y}^{(i)}\right\|^2 \end{aligned} \tag{3.6}$$

where it is the sum of a squared error loss function. To evaluate the linear regression algorithm, it is applied to the ESOL dataset, which is a standard regression dataset with 1128 compounds that estimate the solubility of molecular structures (Delaney, 2004). To implement the linear regression algorithm on the ESOL dataset, we have utilized the scikit-learn package (i.e., sklearn package), which is a free Python package for machine learning. The class **sklearn.linear_model.LinearRegression** creates an object that fits a linear regression model on the given training sample. The overall python code of this part is given in Appendix 1. To evaluate how the learned regression approach performs well on the unseen data, the evaluation measures should be used. In regression, two common evaluation measures are mean squared error (MSE) and r^2. MSE computes the squared sum of the residual between the actual value and the predicted value. Also, r^2 is a statistical measure that computes how well the regressor model fits the real data. The obtained results are shown in Table 3.1. As it is shown, the learned model performs well on training data, but it does worse on the test data. This situation happens frequently in machine learning, which is called overfitting. To cope with this challenge in machine learning, a regularization term is added to the loss function, which controls the size of the parameter vector, and it prevents the parameter vector from becoming large. Hence, the loss function is changed as follows:

$$L\left(\theta; \left\{x^{(i)}, y^{(i)}\right\}_{i=1}^{N}\right) = \sum_{i} \left\| y^{(i)} - h\left(\phi\left(x^{(i)}\right); \theta\right) \right\|^2 = \lambda \sum_{j} |\theta_j|^q \quad (3.7)$$

where λ is a regularization coefficient. If q is set to one, then it is known as Lasso, and if q is set to two, it is known as Ridge loss. The class ***sklearn.linear_model.Ridge*** and ***sklearn.linear_model.Lasso***, respectively, create an object that fits a ridge regression and a lasso regression model on the given training samples. The obtained results by utilizing Ridge regression and Lasso regression on the ESOL dataset are shown in Table 3.1. As it is shown, both of these approaches prevent overfitting and perform better on the ESOL dataset compared to the simple linear regression approach.

To visualize the learned model, 1024 features of each sample are mapped to one dimension using the PCA algorithm. The obtained models on the ESOL dataset using three versions of the linear regression (simple linear regression, linear regression with Ridge regularization, and linear regression with Lasso regularization) are shown in Fig. 3.3. In Fig. 3.3, the red dots show the training samples and the purple dots show the test samples. Also, the blue line shows the predicted line for the training sample.

3.3 Classification

In this section, the classification approaches are studied. Up to now, many classification algorithms are introduced; however, in this section, three main algorithms that perform well in the state-of-the-art are investigated: (1) logistic regression, (2) support vector machine (SVM), and (3) random forest. In the following, each of these algorithms is explained practically.

In this section, to evaluate the classification models, we have used the Tox21 dataset. This dataset is a multi-task dataset that contains 12 tasks (assays) where each task measures the human toxicity of a biological target.

Table 3.1 The obtained results of linear regression on ESOL dataset with three versions of the linear regression (namely, simple linear regression, linear regression with Ridge regularization, and linear regression with Lasso regularization).

	Linear regression				Ridge		Lasso	
	Training data		Test data		Test data		Test data	
Algorithms	MSE		MSE		MSE		MSE	
Without dimension reduction	0.07	0.97	3.26	−1.0	1.13	0.64	2.06	0.27
With dimention reduction with PCA	2.36	0.16	3.96	−0.62	2.48	0.15	2.57	0.11

Both linear regression with Ridge regularization and linear regression with Lasso regularization partly prevent overfitting.

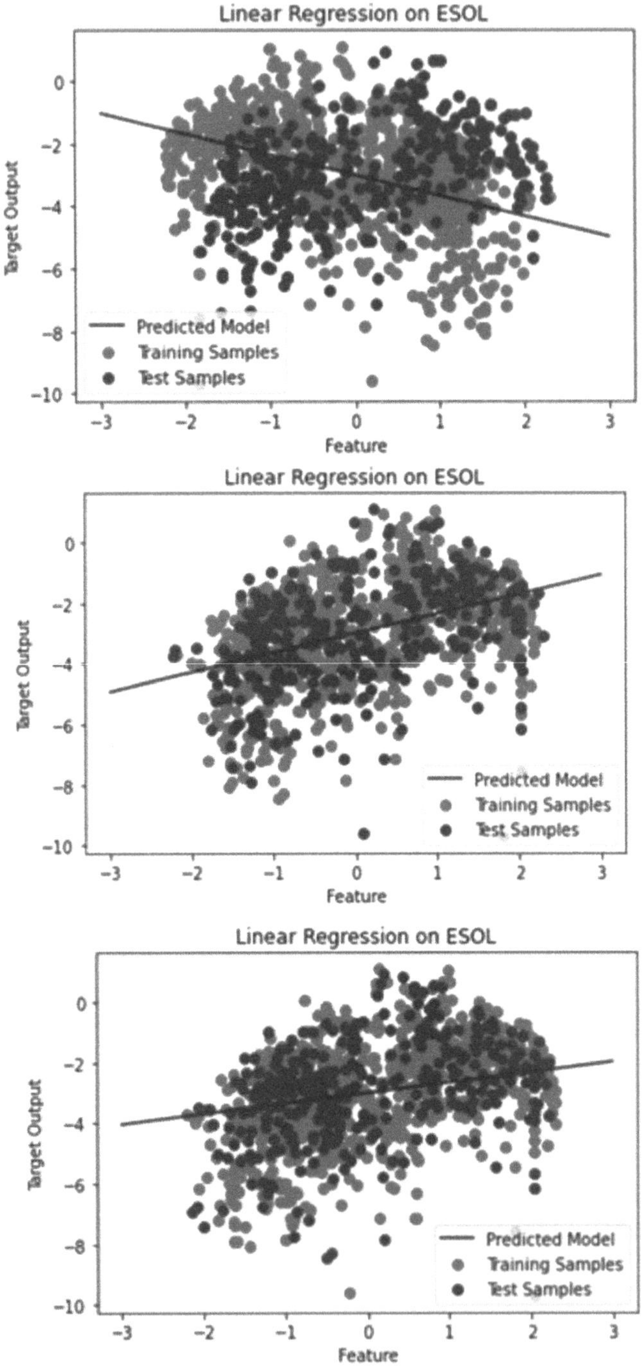

Fig. 3.3 The obtained models on the ESOL dataset using three versions of the linear regression, namely, (A) simple linear regression, (B) linear regression with Ridge regularization, and (C) linear regression with Lasso regularization.

3.4 Logistic regression

Logistic regression is a common binary classifier that aims in discriminating samples between two classes by a hyperplane (see Fig. 3.4), which is called a separator hyperplane. Samples are assigned to their corresponding classes based on which side of the separator hyperplane samples is placed. It should be noted that the multiclass version of the logistic regression also exists.

In logistic regression, the loss function is defined based on the binary cross-entropy, which is as follows:

$$L\left(\theta;\left\{x^{(i)},y^{(i)}\right\}_{i=1}^{N}\right)=\sum_{i}\left[y^{(i)}\log\left(h\left(\phi\left(x^{(i)}\right);\theta\right)\right)+\left(1-y^{(i)}\right)\log\left(1-h\left(\phi\left(x^{(i)}\right);\theta\right)\right)\right] \tag{3.8}$$

where $h\ (\phi(x^{(i)});\theta)$ is defined as follows:

$$h\left(\phi\left(x^{(i)}\right);\theta\right)=\sigma\left(\theta^{T}\phi\left(x^{(i)}\right)\right)=\frac{1}{1+exp\left(-\theta^{T}\phi(x^{(i)})\right)} \tag{3.9}$$

In fact, $h\ (\phi(x^{(i)});\theta)$ encodes the probability of assigning sample $x^{(i)}$ to class one. It should be noted that since $h\ (\phi(x^{(i)});\theta)$ returns a value between

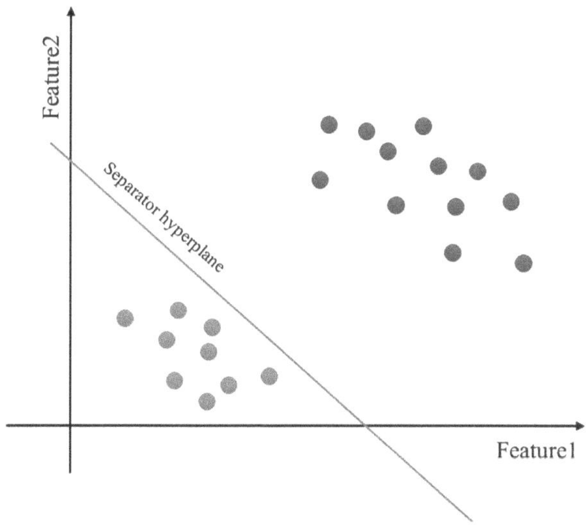

Fig. 3.4 The overall schematic of logistic regression. The *blue (light gray in the printed version) dots* show the samples that belong to class one, and the *green (gray in the printed version)* ones show the samples that belong to class zero. The *orange (dark gray in the printed version) line* shows the separator hyperplane.

zero and one, to assign a label to sample $x^{(i)}$, we need to use a thresholding method. A common choice for threshold value is 0.5; therefore, if $h(\phi(x^{(i)});\theta) \geq 0.5$, the sample is assigned to class one; otherwise, it is assigned to class zero. Hence, to obtain the separator hyperplane, the following equation should be solved:

$$h\left(\phi\left(x^{(i)}\right);\theta\right) = \sigma\left(\theta^T \phi\left(x^{(i)}\right)\right) = 0.5 \tag{3.10}$$

which result in solving the following equation:

$$\theta^T \phi\left(x^{(i)}\right) = 0 \tag{3.11}$$

It should be noted that solving Eq. (3.11) results in a hyperplane in $\phi(x^{(i)})$,which might be a nonlinear separator in space $x^{(i)}$.

To implement the linear regression algorithm on the Tox21 dataset, we have utilized the scikit-learn package. The class ***sklearn.linear_model.LogisticRegression*** creates an object that fits a logistic regression model on the given training sample.

To evaluate the logistic regression model, we have applied it to the Tox21 dataset. In this case, accuracy, F1 measure, and area under curve of ROC (AUC-ROC) are used as evaluation measures. As is mentioned, the Tox21 dataset is a multi-task dataset; hence, in this experiment, we have applied it on three selected random tasks, namely, NR-AR-LBD, NR-AhR, and SR-ARE. The obtained results are shown in Table 3.2.

3.5 SVM

Support vector machine (SVM) is a popular classification algorithm that searches for a separator hyperplane that maximizes the distance of each data point from the separator. The distance between the separator and the closest data point is called margin. Fig. 3.5 shows the obtained separator hyperplane, and it is shown that the distance of the separator hyperplane from the nearest data points is maximum.

SVM formulations are given in two forms: (1) the primal form and (2) the dual form. In the primal form, the SVM loss function is defined as follows:

$$\begin{aligned} &\min_{w,b} \frac{1}{2}\|w\|^2 \\ &s \cdot t \cdot y^{(i)}\left(w^T x^{(i)} + b\right) \geq 1, \quad i = 1, \ldots, m \end{aligned} \tag{3.12}$$

Table 3.2 The obtained results of logistic regression on three selected random tasks of the Tox21 dataset (NR-AR-LBD, NR-AhR, and SR-ARE).

Task	NR-AR-LBD	
Measure	**Training**	**Test**
Accuracy	98.82	95.67
F1-measure	85.35	47.47
AUC-ROC	99.37	76.90
Task	**NR-AhR**	
Measure	**Training**	**Test**
Accuracy	92.78	84.65
F1-measure	76.04	49.57
AUC-ROC	94.89	75.99
Task	**SR-ARE**	
Measure	**Training**	**Test**
Accuracy	87.74	75.45
F1-measure	70.83	42.45
AUC-ROC	89.43	67.66

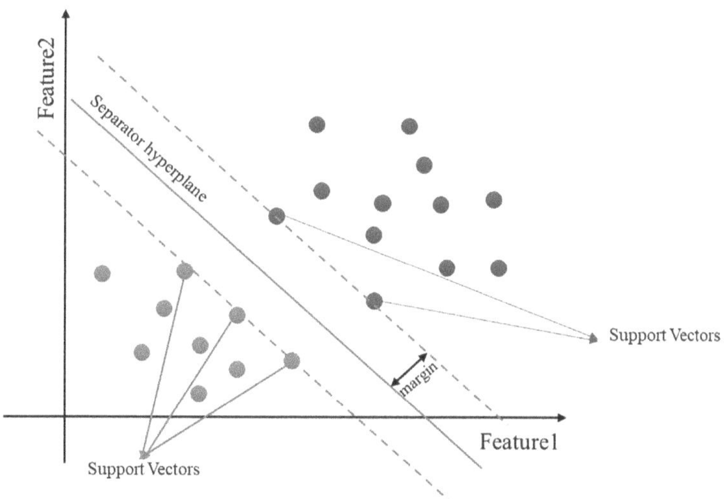

Fig. 3.5 The overall schematic of the SVM classifier. The *blue (light gray in the printed version) dots* show the samples that belong to class one, and the *green (gray in the printed version)* ones show the samples that belong to class zero. The *orange (dark gray in the printed version) solid line* shows the separator hyperplane. Also, the support vectors of each class label are shown. It should be noted that only support vectors are incorporated in the separator hyperplane formation.

where $\frac{1}{2}\ \|w\|^2$ forces the separator hyperplane to have the maximum margin. The constraints (s.t. in formulation) guarantee that all training data should be classified correctly. In the dual form, the following function is considered as loss function:

$$\max_{a} \sum_{i=1}^{m} a_i - \frac{1}{2}\sum_{i=1}^{m} y^{(i)}y^{(j)}a_i a_j K\left(x^{(i)}, x^{(j)}\right)$$
$$s.t.\ a_i \geq 1,\ \ \forall i = 1, \ldots, m,\ \ \sum_{i=1}^{m} a_i y(i) = 0 \tag{3.13}$$

where $K(,.,)$ denotes the kernel function that encodes the similarity between two samples. Also, the kernel function tries to transform the input data into another space in which data is separated by a hyperplane in this new space. There are many common types of kernel functions like polynomial kernel, homogenous polynomial kernel function, and Gaussian RBF kernel function. In Eq. (3.12), each sample has a corresponding α,and this equation states which samples are incorporated in the formation of the separator line. In other words, only samples with nonzero corresponding α value are used in the separator line formation and these samples are called support vectors (see Fig. 3.5).

To implement the SVM algorithm on the Tox21 dataset, we have utilized the scikit-learn package. The class **sklearn.svm.SVC** creates an object that fits a support vector machine model on the given training sample. To evaluate the SVM algorithm, we have applied it to the Tox21 dataset. In this case, two different experiments are done. In the first experiment, it is applied to the NR-AR-LBD task of the Tox21 dataset, which is an imbalanced dataset, since it contains 238 active samples and 6664 inactive samples. In this section, three evaluation measures (namely, accuracy, F1 measure, and AUC-ROC) are used. Also, in this experiment, the kernel function is set to two different values: "rbf" (Gaussian random basis function) and "poly" (polynomial with degree three). This experiment is designed to show how the balancing technique (in imbalanced data) and kernel function impact the final result. The obtained results are shown in Table 3.3. As it is shown, SVM achieves high accuracy but the F1-measure is 0. It means that SVM produces inactive labels for all samples. This challenge is because of the imbalance of the dataset. To cope with this challenge, a weight is assigned to each class label. In other words, a large weight is assigned to a class with a low number of samples. Hence, incorrectly classified samples associated with this class label impose a large penalty in the loss function.

Table 3.3 The obtained results of SVM on NR-AR-LBD task of Tox21 dataset.

Kernel	Task Measure	No balancing technique Training	No balancing technique Test	Use balancing technique Training	Use balancing technique Test
RBF	Accuracy	96.69	96.23	96.83	96.96
	F1-measure	0.0	0.0	58.76	60.87
	AUC-ROC	50.00	50.00	83.38	79.82
	Number of support vectors	424		2692	
Polynomial	Accuracy	96.42	96.86	3.35	3.67
	F1-measure	0.0	0.0	6.49	7.08
	AUC-ROC	50.00	50.00	50.00	50.00
	Number of support vectors	346		4831	

This experiment shows how the balancing technique (in imbalanced data) and kernel function impact the final result.

To take into account this issue, in-class ***sklearn.svm.SVC,*** an argument "class_weight" is considered, which could be set manually or could be set to "balanced." In this experiment, it is set to "balanced." The Python code of this experiment is given in Appendix. The obtained results are given in Table 3.3. As it is shown, a reliable model is learned using the balancing technique. In Table 3.3, the number of support vectors is also reported. A learned model with a low number of support vectors gets the better value in all three evaluation measures. It should be noted that a model with a large number of support vectors means that all training samples are placed on margins or between margins.

As it is mentioned, the Tox21 dataset is a multi-task dataset. In the second experiment, we have applied it to three random tasks, namely, NR-AR-LBD, NR-AhR, and SR-ARE. This experiment is designed to show how SVM performs on three different tasks. It should be noted that in this experiment, the kernel function is set to "rbf" and the balancing technique is utilized by setting class_weight to "balanced." The obtained results are shown in Table 3.4. As it is shown, SVM performs better on the NR-AR-LBD task than on the two other tasks.

3.6 Random forest

Random forest is a state-of-the-art machine learning method, which in many research areas achieves comparable performance with respect to the

Table 3.4 The obtained results of SVM on three selected random tasks of the Tox21 dataset (namely, NR-AR-LBD, NR-AhR, and SR-ARE).

Task	NR-AR-LBD	
Measure	Training	Test
Accuracy	97.25	96.75
F1-measure	62.74	55.10
AUC-ROC	82.74	77.93
Task	NR-AhR	
Measure	Training	Test
Accuracy	73.24	72.04
F1-measure	43.92	41.64
AUC-ROC	80.20	77.87
Task	SR-ARE	
Measure	Training	Test
Accuracy	66.85	65.22
F1-measure	41.63	39.12
AUC-ROC	69.32	66.72

The kernel function is set to "rbf," and the balancing technique is utilized.

other methods. Random forest is an ensemble learning method for regression, classification, and other tasks in which diverse decision trees as base predictors are generated. The final decision is made using voting (in classification), mean on average prediction of each tree (in regression), or other fusion techniques. The decision tree is a tree-like structure in which each internal node makes a decision about a feature that is effective in the task prediction process. Also, a class label is assigned to each leaf node. To classify an unseen input, the tree is traversed from root to leaf using input features to assign an output to it.

In the decision tree, there are two main challenges: how to split nodes and how to find the best trees. It is shown that learning the simplest decision tree is an NP-complete problem (Hyafil & Rivest, 1973). To this end, a greedy heuristic is used to construct a decision tree. At first, the algorithm is started from an empty decision tree, and then the next best feature is selected to split using some measures like information gain (Quinlan, 1986), Gini impurity, or variance reduction (Breiman, Friedman, Olshen, & Stone, 1984). This procedure is iterated until the tree is constructed. Now, there are many variants of the decision tree algorithm like ID3, C4.5, and CART algorithm (Mitchell, 1997).

To implement a random forest algorithm in Python, scikit-learn package could be used. The class ***sklearn.ensemble.RandomForestClassifier*** creates an object that fits a random forest classifier model on the given training samples. This class has many parameters like "*n_estimator*" denotes the number of decision trees in a random forest, "*criterion*" denotes the measure to select the next feature to split, and "*max_depth*" denotes the maximum depth of each decision tree.

To evaluate random forest on Tox21, a similar experiment like the two previous sections is designed and it is applied on three randomly selected tasks of Tox21. The obtained results are shown in Table 3.5. As it is shown, random forest, like SVM, performs better on NR-AR-LBD tasks than on the two other tasks.

A property called "*feature_importances*" in ***sklearn.ensemble.RandomForestClassifier*** object assigns a weight to each feature, which denotes how important a feature is in prediction. Fig. 3.6 shows the "*feature_importances*"

Table 3.5 The obtained results of random forest on three selected random tasks of the Tox21 dataset, namely, NR-AR-LBD, NR-AhR, and SR-ARE.

Task	NR-AR-LBD	
Measure	**Training**	**Test**
Accuracy	99.93	98.04
F1-measure	98.96	65.20
AUC-ROC	99.91	76.58
Task	**NR-AhR**	
Measure	**Training**	**Test**
Accuracy	99.82	90.63
F1-measure	99.24	44.91
AUC-ROC	99.86	65.54
Task	**SR-ARE**	
Measure	**Training**	**Test**
Accuracy	66.85	65.22
F1-measure	41.63	39.12
AUC-ROC	69.32	66.72

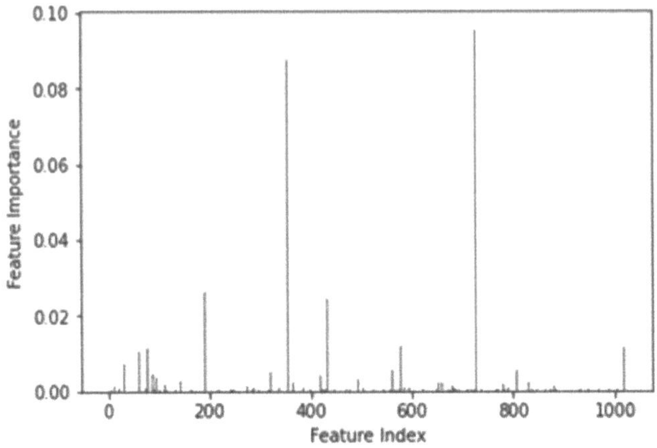

Fig. 3.6 The feature importance of the learned random forest model on the NR-AhR task.

property for the learned random forest model. As it is shown, the 726th feature gets the higher importance than the other features on the learned random forest model on the NR-AhR task.

3.7 Deep learning-based models

In this section, deep learning-based models for classification and regression are investigated. In these methods, both feature extraction and task prediction tasks are done automatically. In other words, for each input sample, based on the task, a custom feature vector is learned. To this end, in the following, the first classic neural network is explained, and then the convolutional neural network is explained practically.

Hinton, Osindero, and Teh (2006) introduced an approach in which a learning algorithm for deep belief nets is given. One of the most important contributions of their work was the unsupervised learning of each hidden layer. After this step, a whole network is learned by utilizing the supervised knowledge. Actually, in Hinton et al. (2006), features are automatically learned. It gives rise to deep learning to automatically learn features from input data hierarchically. In recent years, more deep learning-based networks are introduced.

3.8 Classic neural network

The classic neural network consists of multiple layers (input layer, some hidden layers, and output layer) where each layer itself consists of several

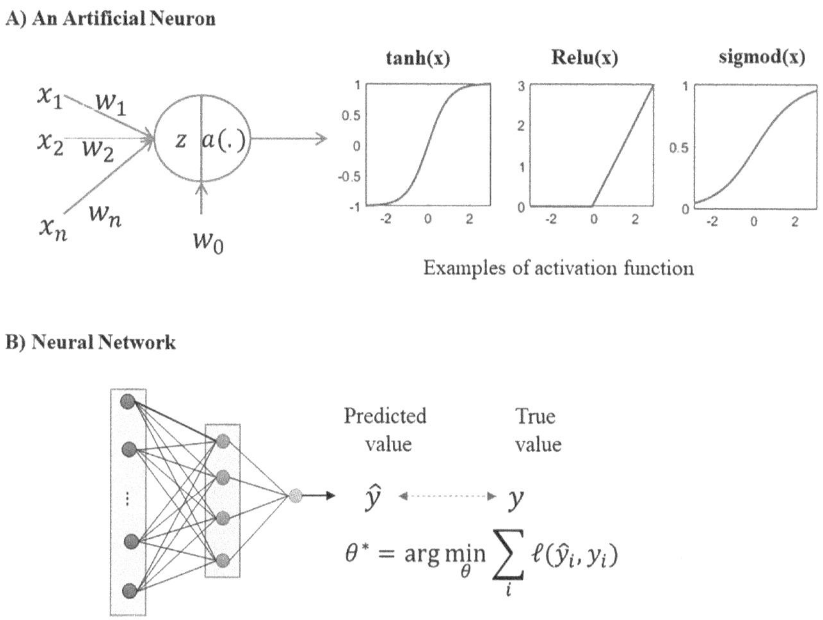

Fig. 3.7 Learning the parameters of the model is done by optimizing the loss function optimization. (A) A schematic view of an artificial neuron. (B) A simple neural network with an input layer, one hidden layer, and an output layer.

artificial neurons. The structure of each artificial neuron is inspired by human brain biological neurons. As it is represented in Fig. 3.7, each neuron has an input vector and a scalar as an output. In this case, at first, the weighted sum of the input is computed and then its output is fed into the nonlinear activation function. There are many knowns of the activation functions like sigmoid function, rectified linear unit (ReLU), and softmax function. However, any nonlinear differentiable function could be used (Abbasi et al., 2021).

To design a classic neural network, its structure including the number of layers, the number of neurons in each layer, and the activation functions of each layer should be determined. Then, its loss function and the optimizer approach of the loss function are defined. Assume, we want to use the neural network for a regression problem. In this case, the architecture contains the input layer, two hidden layers with the Relu activation function, and one output layer with a linear activation function. Also, the loss function is mean squared error and the optimizer is stochastic gradient descent (SGD). To implement this network, the Keras package with Tensorflow backend is utilized. The code of this part is shown in Appendix 1. To evaluate this

Table 3.6 The obtained results of the classic neural network are reported on the ESOL dataset.

	Training data		Test data	
Algorithms	MSE		MSE	
Classic neural network	0.94	0.67	1.73	0.38
Classic neural network with ridge regularization	1.11	0.61	1.71	0.40
Classic neural network with Lasso regularization	1.09	0.61	1.69	0.41

Also, the results of classic neural networks with two regularizations, namely, Ridge and Lasso, are reported.

network, it is applied to the ESOL dataset. The obtained results are shown in Table 3.6. It should be noted that there are many hyperparameters that should be set like the learning rate, batch size, and the number of epochs that are so effective in the final result. The obtained results of the classic neural network, the classic neural network with Ridge regularization, and the classic neural network with Lasso regularization are reported on the ESOL dataset. As it is shown, the classic neural network with Lasso regularization performs better than the other two ones.

To investigate how neural network performs on the classification task, three random tasks, similar to the selected tasks in section classic approaches, namely, NR-AR-LBD, NR-AhR, and SR-ARE, are chosen. The architecture of this network in the classification task is similar to that of the regression task with the difference that the activation function of the last layer is sigmoid. Also, in this experiment, one-hot encoding of the SMILES is fed as input to the network. Table 3.7 shows the obtained results. In this case, the neural network receives acceptable performance on three tasks.

3.9 Convolutional neural network

In this section, convolutional neural networks, also called ConvNets, are investigated. ConvNets was first introduced by LeCun et al. (1989) in the 1980s. Then, the first ConvNet, called LeNet, is introduced, which recognizes handwritten digits. However, CNNs do not get attention for several years due to their high computational complexity and low power of computer resources in that decade. Krizhevsky, Sutskever, and Hinton (2012) introduced a CNN for image classification, called AlexNet, which achieves high accuracy in ImageNet Large Scale Visual Recognition Challenge compared to other approaches. It causes CNNs to be revisited, and a new field of research called deep learning gets more attention.

Table 3.7 The obtained results of the neural network on three tasks of the Tox21 dataset.

Task Measure	NR-AR-LBD Training	 Test
Accuracy	99.19	97.64
F1-measure	91.32	65.36
AUC-ROC	99.93	90.51
Task Measure	**NR-AhR Training**	** Test**
Accuracy	98.85	91.69
F1-measure	95.58	67.75
AUC-ROC	99.92	90.17
Task Measure	**SR-ARE Training**	** Test**
Accuracy	98.29	88.29
F1-measure	95.14	64.59
AUC-ROC	99.89	85.72

CNN is a network that commonly consists of four types of layers: convolutional layer, pooling layer, normalization layer, and dropout layer. In the following, each of these layers is explained, and then a practical example of how to use CNN in drug molecular property prediction is explained.

Convolutional layer: this layer is based on the convolutional operator in signal processing. The input of this layer could be 1D, 2D, or 3D signal. Each layer has n filters, and the size of the filters is determined based on the input, which is 1D, 2D, or 3D signal. It should be noted that the content of these filters are parameters and during the training, the procedure should be learned. In this layer, the filter is placed in the center of each element of the input signal and the content of the filter and the input signal are element-wisely multiplied. Then, the filter is slid over the input signal to do these operations on all elements. The direction of the sliding is dependent on the input signal. For 1D signal, sliding is done only in one direction, for 2D signal, sliding is done in two directions, and so on (see Fig. 3.5). The convolutional layer semantically extracts the structured local features. By stacking the convolutional layers, mid-level and high-level features could be extracted.

Pooling layer: the main goal of this layer is to increase the field of view by using a downsampling operation. The pooling layer is typically placed after a convolution layer. Two common types of this layer are the max-pooling layer and the average pooling layer. In the max-pooling layer, the maximum value of the local filter is taken, and in the average pooling layer, the average value of the local filter is computed. It should be noted that the pooling layer has no learnable parameters.

By stacking the convolutional layers, the field of view is increased; however, in this case, many learnable filters are defined. By utilizing the pooling layer, we could increase the field of view without defining new parameters.

Normalization layer: this layer normalizes the output of the previous layer before feeding it into the next layer. There are two common normalization layers, namely, batch normalization layer (Ioffe & Szegedy, 2015) and layer normalization (Ba, Kiros, & Hinton, 2016). Both of these layers adaptively normalize the layer and have their own mean and variance. In the batch normalization layer, a mean and variance for normalization are computed over the minibatch, while in the normalization layer, the mean and variance used are computed from all of the summed inputs to the neurons in a layer on a single training case. It causes normalization layer, unlike batch normalization, to perform the same computation in both training and test times.

Dropout layer: this layer is designed such that it helps the model prevent overfitting (Srivastava, Hinton, Krizhevsky, Sutskever, & Salakhutdinov, 2014). It leads the model to have more generalization power. This layer has no learnable parameters. This layer sets some randomly selected elements of the input to zero, and the other remaining ones are transferred to the output without change.

CNN is an architecture comprising the four mentioned layers (custom layers could be defined). The CNN architecture is custom-determined by a designer. In this section, to evaluate how CNN performs in molecule property prediction, some experiments are designed. In this case, two tasks are chosen: classification and regression. In both cases, the architecture of CNN contains three convolutional layers, a max-pooling layer, one convolutional layer, a fully connected layer, a dropout layer, a fully connected layer, a dropout layer, and two fully connected layers. In the regression task, the mean squared error is considered a loss function. In the classification task, the loss function is binary cross-entropy. In both cases, the number of epochs

Table 3.8 The obtained results of the classic neural network are reported on the ESOL dataset.

	Training data		Test data	
Algorithms	**MSE**		**MSE**	
Convolutional neural network	0.44	0.84	0.58	0.80
Convolutional neural network with ridge regularization	0.34	0.88	0.42	0.84
Convolutional neural network with Lasso regularization	0.42	0.86	0.54	0.83

Also, the results of classic neural networks with two regularizations, namely, Ridge and Lasso, are reported.

is set to 100 and the number of batch size is set to 256. The obtained results on the regression task are shown in Table 3.8. As it is represented, CNN with ridge regularization gets the lowest MSE and highest r^2 measure. The classification task results are given in Table 3.9. CNN performs best on the three tasks in the training phase. Also, CNN has received reliable results in the test phase.

Table 3.9 The obtained results of the convolutional neural network on three tasks of the Tox21 dataset.

Task	**NR-AR-LBD**	
Measure	**Training**	**Test**
Accuracy	100	98.70
F1-measure	100	78.76
AUC-ROC	100	87.39
Task	**NR-AhR**	
Measure	**Training**	**Test**
Accuracy	100	95.05
F1-measure	100	79.11
AUC-ROC	100	91.21
Task	**SR-ARE**	
Measure	**Training**	**Test**
Accuracy	100	92.13
F1-measure	100	73.62
AUC-ROC	100	87.90

4. Conclusion

In this chapter, multivariate machine learning algorithms are practically investigated. To do so, these algorithms, based on the feature extraction step, are divided into two categories: the classic approaches and the deep learning-based approaches. In both of these categories, regression and classification tasks are studied. In regression, the introduced approaches are applied to the ESOL dataset. Three regression methods are investigated: linear regression, neural network, and convolutional neural network. To compare how these methods perform on the ESOL dataset, the obtained results of these methods of the regression task are summarized in Table 3.10. The results show that CNN with ridge regularization performs better than the other comparable methods. It should be noted that the high r^2 value with a low MSE value is acceptable.

The classification task is also investigated. To this end, three classic approaches (namely, logistic regression, SVM, and random forest) and two deep learning-based approaches (namely, neural network and convolutional neural network) are studied. Similar to the regression task, the obtained results of the classification task on NR-AR-LBD, NR-AhR, and SR-ARE tasks are, respectively, shown in Tables 3.11–3.13. In all cases,

Table 3.10 Comparing results of the investigated regression methods on the ESOL dataset.

Category	Algorithms	Test data	
		MSE	
Classic approaches	Linear regression	3.26	−1.0
	Linear regression with Ridge regularization	1.13	0.64
	Linear regression with Lasso regularization	2.06	0.27
Deep learning-based approaches	Neural network	1.73	0.38
	Neural network with Ridge regularization	1.71	0.40
	Neural network with Lasso regularization	1.69	0.41
	Convolutional neural network	0.58	0.80
	Convolutional neural network with Ridge regularization	0.42	0.84
	Convolutional neural network with Lasso regularization	0.54	0.83

Table 3.11 Comparing results of the investigated classification methods on NR-AR-LBD task of Tox21 dataset.

		Measures		
Category	**Algorithms**	**Acc.**	**F1**	**AUC-ROC**
Classic approaches	Logistic regression	95.67	47.47	76.90
	SVM	96.75	55.10	77.93
	Random forest	98.04	65.20	76.58
Deep learning-based approaches	Neural network	97.64	65.36	90.51
	Convolutional neural network	98.70	78.76	87.39

Table 3.12 Comparing results of the investigated classification methods on NR-AhR task of Tox21 dataset.

		Measures		
Category	**Algorithms**	**Acc.**	**F1**	**AUC-ROC**
Classic approaches	Logistic regression	90.63	44.91	65.54
	SVM	72.04	41.64	77.87
	Random forest	84.65	49.57	75.99
Deep learning-based approaches	Neural network	91.69	67.75	90.17
	Convolutional neural network	95.05	79.11	91.21

Table 3.13 Comparing results of the investigated classification methods on SR-ARE task of Tox21 dataset.

		Measures		
Category	**Algorithms**	**Acc.**	**F1**	**AUC-ROC**
Classic approaches	Logistic regression	75.45	42.45	67.66
	SVM	65.22	39.12	66.72
	Random forest	65.22	39.12	66,372
Deep learning-based approaches	Neural network	88.29	64.59	85.72
	Convolutional neural network	92.13	73.62	87.9

CNN performs better than the other approaches. It should be noted that all three tasks are imbalanced; nevertheless, CNN in F1-measure and AUC-ROC get acceptable results. It is due to automatically feature extraction step. In classic approaches, ECFP for the input molecules is computed and then it feeds into the model. However, in neural network and convolutional neural network, one-hot encoding of the input molecule is fed into the network and the feature vector is automatically learned.

4.1 Advanced topics

In deep learning, CNNs are investigated more, and some architectures for many tasks are given (He, Zhang, Ren, & Sun, 2015; Krizhevsky et al., 2012; Szegedy et al., 2015). Also, other new concepts are introduced, which act successfully in many research areas, including generative adversarial network (GAN) (Goodfellow et al., 2014), capsule network (Hinton, Sabour, & Frosst, 2018; Sabour, Frosst, & Hinton, 2017), graph representational learning (Fey & Lenssen, 2019; Hamilton, 2020; Wang et al., 2018), and deep reinforcement learning (Henderson et al., 2018). GAN is an unsupervised method that learns the distribution of the input space. GAN contains two subnetworks: generator and discriminator; these two subnetworks learn throughout the adversarial learning approach. Hence, it could generate a new sample from the learned generator. The capsule network is designed such that it addresses the most drawback of the CNN. CNN could not take into account important spatial hierarchies between features. In computational biology, there is more situation in which the input space is represented by graphs. Hence, graph representational learning methods to encode the knowledge play a more important role. Deep reinforcement learning is a field of machine learning whose main goal is to predict the policy function (decision-making function) by interacting with the environment. In this case, the agent is encouraged or punished by a reward function based on the selected action.

These concepts get more attention in bioinformatics and computational biology. Yang, Chen, Li, and Wang (2021) introduced the subtype-GAN model with multiple-input and multiple-output to identify tumor subtypes. To this end, they have utilized a deep adversarial learning approach. Lan et al. (2020) investigated the applications of GAN in biomedical informatics. Popova, Isayev, and Tropsha (2018) proposed a new approach for Denovo drug design, which is called ReLeaSE (Reinforcement Learning for Structural Evolution). They want to generate novel molecules libraries. Also, some research works utilize the capsule network for different problems in computational biology (de Jesus, Cuevas, Rivera, & Crivelli, 2018; Liu et al., 2020; Mobiny, Lu, Nguyen, Roysam, & Varadarajan, 2019).

References

Abbasi, K., Razzaghi, P., Poso, A., Ghanbari-Ara, S., & Masoudi-Nejad, A. (2021). Deep learning in drug target interaction prediction: Current and future perspectives. *Current Medicinal Chemistry*, *28*(11), 2100–2113.

Ba, J. L., Kiros, J. R., & Hinton, G. E. (2016). Layer normalization. *arXiv Preprint arXiv:1607.06450.*

Bishop, C. M. (2006). *Machine learning and pattern recognition*. Heidelberg: Springer.

Breiman, L., Friedman, J. H., Olshen, R. A., & Stone, C. J. (1984). Classification and regression trees. In *Breiman classification and regression trees*. Pacific Grove, CA: Wadsworth and Brooks.

David, L., Thakkar, A., Mercado, R., & Engkvist, O. (2020). Molecular representations in AI-driven drug discovery: A review and practical guide. *Journal of Cheminformatics*, *12*(1), 1–22.

de Jesus, D. R., Cuevas, J., Rivera, W., & Crivelli, S. (2018). Capsule networks for protein structure classification and prediction. *arXiv Preprint arXiv:1808.07475.*

Delaney, J. S. (2004). ESOL: Estimating aqueous solubility directly from molecular structure. *Journal of Chemical Information and Computer Sciences*, *44*(3), 1000–1005.

Durant, J. L., Leland, B. A., Henry, D. R., & Nourse, J. G. (2002). Reoptimization of MDL keys for use in drug discovery. *Journal of Chemical Information and Computer Sciences*, *42*(6), 1273–1280.

Fey, M., & Lenssen, J. E. (2019). Fast graph representation learning with PyTorch Geometric. *arXiv Preprint arXiv:1903.02428.*

Gerdes, H., Casado, P., Dokal, A., Hijazi, M., Akhtar, N., Osuntola, R., et al. (2021). Drug ranking using machine learning systematically predicts the efficacy of anti-cancer drugs. *Nature Communications*, *12*(1), 1–15.

Goodfellow, I., Pouget-Abadie, J., Mirza, M., Xu, B., Warde-Farley, D., Ozair, S., et al. (2014). Generative adversarial nets. In *Vol. 27. Advances in neural information processing systems.*

Hamilton, W. L. (2020). Graph representation learning. *Synthesis Lectures on Artifical Intelligence and Machine Learning*, *14*(3), 1–159.

He, K., Zhang, X., Ren, S., & Sun, J. (2015). Deep residual learning for image recognition. *arXiv:1512.03385.*

Henderson, P., Islam, R., Bachman, P., Pineau, J., Precup, D., & Meger, D. (2018). Deep reinforcement learning that matters. In *Vol. 32. AAAI conference on artificial intelligence.*

Hinton, G. E., Osindero, S., & Teh, Y. W. (2006). A fast learning algorithm for deep belief nets. *Neural Computation*, *18*(7), 1527–1554.

Hinton, G. E., Sabour, S., & Frosst, N. (2018). Matrix capsules with EM routing. In *International conference on learning representations.*

Hyafil, L., & Rivest, R. L. (1973). *Laboratoire de recherche.*

Ioffe, S., & Szegedy, C. (2015). Batch normalization: Accelerating deep network training by reducing internal covariate shift. In *International conference on machine learning* (pp. 448–456).

Jumper, J., Evans, R., Pritzel, A., Green, T., Figurnov, M., Ronneberger, O., et al. (2021). Highly accurate protein structure prediction with AlphaFold. *Nature*, *596*(7873), 583–589.

Krizhevsky, A., Sutskever, I., & Hinton, G. E. (2012). Imagenet classification with deep convolutional neural networks. In *Vol. 25. Advances in neural information processing systems* (pp. 1097–1105).

Lan, L., You, L., Zhang, Z., Fan, Z., Zhao, W., Zeng, N., et al. (2020). Generative adversarial networks and its applications in biomedical informatics. *Frontiers in Public Health*, *8*, 164.

LeCun, Y., Boser, B., Denker, J. S., Henderson, D., Howard, R. E., Hubbard, W., et al. (1989). Backpropagation applied to handwritten zip code recognition. *Neural Computation, 1*(4), 541–551.

Liu, Y., Ding, Y., Li, C., Cheng, J., Song, R., Wan, F., et al. (2020). Multi-channel EEG-based emotion recognition via a multi-level features guided capsule network. *Computers in Biology and Medicine, 123*, 103927.

Mitchell, T. (1997). *Machine learning*. McGraw Hill.

Mobiny, A., Lu, H., Nguyen, H. V., Roysam, B., & Varadarajan, N. (2019). Automated classification of apoptosis in phase contrast microscopy using capsule network. *IEEE Transactions on Medical Imaging, 39*(1), 1–10.

Popova, M., Isayev, O., & Tropsha, A. (2018). Deep reinforcement learning for de novo drug design. *Science Advances, 4*(7), eaap7885.

Quinlan, J. R. (1986). Induction of decision trees. *Machine Learning, 1*(1), 81–106.

Rogers, D., & Hahn, M. (2010). Extended-connectivity fingerprints. *Journal of Chemical Information and Modeling, 50*(5), 742–754.

Sabour, S., Frosst, N., & Hinton, G. E. (2017). Dynamic routing between capsules. *arXiv Preprint arXiv:1710.09829*.

Srivastava, N., Hinton, G., Krizhevsky, A., Sutskever, I., & Salakhutdinov, R. (2014). Dropout: A simple way to prevent neural networks from overfitting. *Journal of Machine Learning Research, 15*(1), 1929–1958.

Szegedy, C., Liu, W., Jia, Y., Sermanet, P., Reed, S., Anguelov, D., et al. (2015). Going deeper with convolutions. In *IEEE Conference on Computer Vision and Pattern Recognition* (pp. 1–9).

Wang, H., Wang, J., Wang, J., Zhao, M., Zhang, W., Zhang, F., et al. (2018). Graphgan: Graph representation learning with generative adversarial nets. *AAAI Conference on Artificial Intelligence, 32*(1).

Yang, H., Chen, R., Li, D., & Wang, Z. (2021). Subtype-GAN: A deep learning approach for integrative cancer subtyping of multi-omics data. *Bioinformatics, 37*(16), 2231–2237.

Yang, C., Tarkhov, A., Marusczyk, J., Bienfait, B., Gasteiger, J., Kleinoeder, T., et al. (2015). New publicly available chemical query language, CSRML, to support chemotype representations for application to data mining and modeling. *Journal of Chemical Information and Modeling, 55*(3), 510–528.

CHAPTER 4

Tuning the apparent thermodynamic parameters of chemical systems

Hamid Abdollahi[a], Somaiyeh Khodadadi Karimvand[b], and Somaye Vali Zade[c]
[a]Department of Chemistry, Institute for Advanced Studies in Basic Sciences (IASBS), Zanjan, Iran
[b]Department of Chemistry, Institute for Advanced Studies in Basic Sciences, Zanjan, Iran
[c]Halal Research Center of IRI, Food and Drug Administration, Ministry of Health and Medical Education, Tehran, Iran

1. Introduction

For many important equilibrium systems, it is customary to use a simplified apparent, instead of a complete but rather complex, equilibrium system. The resulting apparent equilibrium constants are altered or controlled by changes in the chemical system, usually modifying the total concentration of auxiliary reagents, but the variation in other parameters such the intensity of light in photo-controlled reactions, types and concentration of solvents or composition of mixed solvents, the ionic strength of the chemical media, etc. is also possible.

It is common to use the conditional formation constant, or the effective formation constant in the interactions between metal ions and protons of synthetic chelating agents such as EDTA, EGTA, or with pH indicators. For example, the following equilibria exhaustively describe all interactions between metal ion Ca^{2+} and with pH indicator Murexide (Mur) in aqueous solution (Ringbom, 1963) (Fig. 4.1).

The collection of 8 equilibrium constants (K_{a1}, K_{a2}, K_{a3}, K_{a4}, K_{a5}, K_1, K_2, and K_3) together with the total concentrations for the components quantitatively defines the speciation in any solution of the three components. For the most relevant applications, i.e., complexometric titrations, this complex equilibrium system is reduced to an apparent simple equilibrium under the condition of constant pH. Most titrations are conducted at a certain pH in buffered solutions.

$$Ca^{2+} + Mur \rightleftharpoons CaMur \quad K' = \frac{[CaMur]}{[Ca^{2+}][Mur']} \tag{4.1}$$

Machine Learning and Pattern Recognition Methods in Chemistry from Multivariate and Data Driven Modeling
https://doi.org/10.1016/B978-0-323-90408-7.00008-3

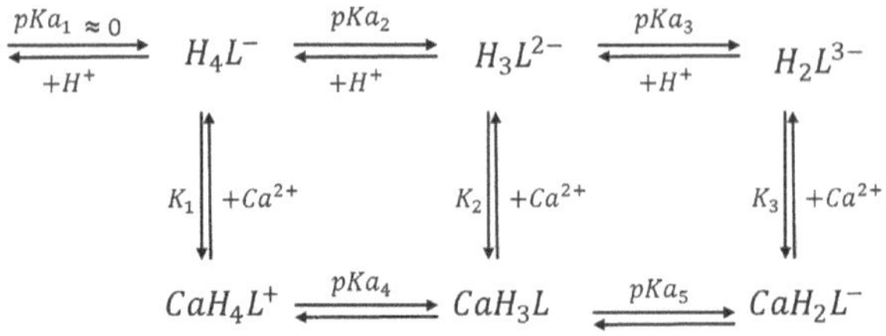

Fig. 4.1 The equilibrium reactions between Ca^{2+} ions and Murexide in aqueous solution.

K′ is the conditional formation constant, which describes the action quantitatively as long as the pH is constant (Tammelin & Mogensen, 1952), and its value is pH dependent. In other words, there is a range of the apparent equilibrium constant, K′, which can be tuned by adjusting the pH in a predefined value. However, the color of solution of indicator is dependent on two factors: the kind of metal ion and the pH of solution.

The concept of equilibrium reactions is extremely important in chemistry as many main applications of chemistry processes are based on the laws that govern the equilibrium reactions. In a typical equilibrium reaction, the amount of activities of involved components is determined by the relevant thermodynamic constant. As the activities of involved components in the equilibrium reaction are controlled by the thermodynamic parameters, tuning the relevant equilibrium constants can control the efficiency of the equilibria-based process like chemical sensing. The main goal of this chapter is to quantitatively investigate the effects of side equilibria in order to find the tunable range of the apparent or conditional constants of different chemical equilibrium systems.

2. General strategy for tuning the apparent constant (Karimvand, Maeder, & Abdollahi, 2019)

To further investigate the issues of using apparent equilibrium constants, we quantitatively analyze a simpler equilibrium system.

$$\mathrm{A} + \mathrm{B} \rightleftharpoons \mathrm{C}^{K_C} = \frac{[\mathrm{C}]}{[\mathrm{A}][\mathrm{B}]} \tag{4.2}$$

Consider a situation where both the reactant and the product are involved in competitive equilibrium reactions with an interferent I, which is represented in Fig. 4.2.

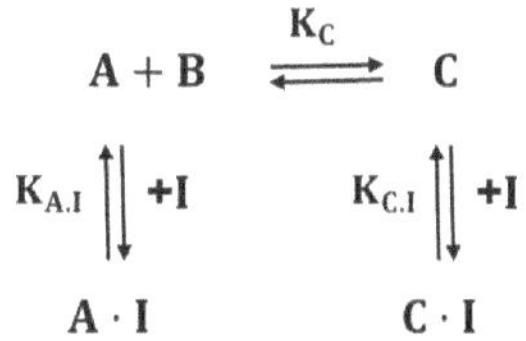

Fig. 4.2 The basic equilibrium A+B=C. The basic equilibrium A+B=C, with two auxiliary equilibrium interactions between the interferent I and the equilibrium species A and C.

In addition to the initial equilibrium, there are now two additional interactions of I with A and with C:

$$K_{A.I} = \frac{[A.I]}{[A][I]} \tag{4.3}$$

$$K_{C.I} = \frac{[C.I]}{[C][I]} \tag{4.4}$$

The apparent equilibrium reaction and its constant $K'_{C'}$ are defined as:

$$A' + B \rightleftharpoons C' \; K'_{C'} = \frac{[C']}{[A'][B]} \tag{4.5}$$

The hypothetical components A' and C' are defined as the sum over all species containing component A or C.

$$[A'] = [A] + [A.I] \tag{4.6}$$

And:

$$[C'] = [C] + [C.I] \tag{4.7}$$

Thus, in the presence of the side reaction(s), the relationship between the hypothetical components A' and C' and the real species of the system can be defined by combining Eqs. (4.5)–(4.7):

$$K'_{C'} = \frac{[C] + [C.I]}{([A] + [A.I])[B]}$$

Next, Eqs. (4.3) and (4.4) are used to replace [A.I] and [C.I]

$$K'_{C'} = \frac{[C] + K_{C.I}[C][I]}{([A] + K_{A.I}[A][I])[B]}$$

Rewriting results in

$$K'_{C'} = \frac{[C]}{[B][A]} \times \frac{1 + K_{C.I}[I]}{1 + K_{A.I}[I]}$$

Therefore,

$$K'_{C'} = K_C \frac{1 + K_{C.I}[I]}{1 + K_{A.I}[I]} \tag{4.8}$$

In this system of complete and apparent reaction schemes, an analytical relationship between the real and the apparent constants can be established. According to Eq. (4.8), the value of the apparent constant is defined by the constants of the complete model K_C, $K_{A.I}$, and $K_{C.I}$ as well as the free [I] concentration of the interferent I. Eq. (4.8) also shows that the influence of [I] on $K'_{C'}$ is more prominent if the equilibrium constants $K_{A.I}$ and $K_{C.I}$ are dissimilar. If the interferent is added in large excess, its free concentration is approximately constant and thus also the value for the apparent equilibrium constant $K'_{C'}$. If just the reactant A gets involved in the side reaction, $K_{C.I}$ would be zero, and in the situation where only the product (C) is involved in the side reaction, $K_{A.I}$ is zero.

Now, this question can be asked that under what situation, the apparent model and the related constant ($K'_{C'}$) are applicable in an equilibrium system. By involving the side reaction(s), total concentrations of all components can be applied as controllable variables for changing [I]. Thus, Eq. (4.5) is applicable, where [I] is independent of the total concentrations of all involved components (A, B, and I).

The question is, "under what circumstances can the apparent model and the related constant ($K'_{C'}$) be applied to an equilibrium system?" By involving the side reaction(s), the total concentrations of all components will influence [I]. Thus, Eqs. (4.5) and (4.8) are applicable where [I] is independent of the total concentrations of all involved components (A, B, and I). In this section, we investigate situations where the apparent model is valid.

Herein, the effect of total concentration of interferent I (which is introduced to the system and can be used to tune the apparent constant) on [I] and as a result, the stability of apparent constant during a typical titration is investigated. As an example, in Fig. 4.3, the value of [I] as a function of C_I during the titration for $K_C = 10^6$, $K_{A.I} = 10^{4.5}$, and $K_{C.I} = 10^{1.2}$ has been shown. It is clearly manifested from Fig. 4.3 that in the low total concentrations of the interferent I, the free concentration of I, [I], and thus the apparent constant are not independent of the equilibrium situation (Part A in Fig. 4.3), while part B shows the region where C_I and [I] are independent. Therefore, since [I] in each titration is approximately constant in a certain range of total concentrations of I, (the percent change of [I] during the system under study is $\leq 1\%$), this range of C_I can be applicable for applying the apparent model.

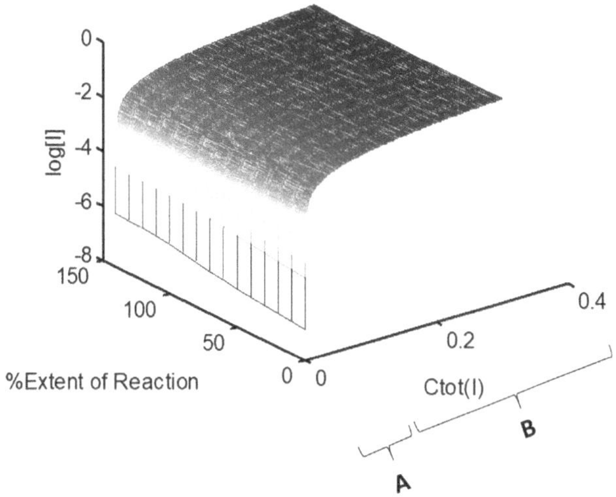

Fig. 4.3 Plot for log of [I] against different concentrations of entered component I for the titration of 10 mL A ($C_A = 1 \times 10^{-4}$) with 0.14 mL B ($C_B = 1 \times 10^{-2}$) when C_I changes from 1×10^{-6} to 3×10^{-1}.

Notably, this range is practically limited in the higher concentrations of I. Thus, for any equilibrium system, a limited range of interferent concentration can be achieved in which the apparent model is applicable.

As can be seen clearly from Fig. 4.3, in the permissible range of the apparent model (part B), by changing the total concentration of I (C_I), [I] is altered. Thus, in part B, by changing C_I, the apparent constant will be changed. Obviously, this is the permissible range of tuning the apparent constant. Fig. 4.4 shows the dependency of apparent constant ($K'_{C'}$) in its calculated

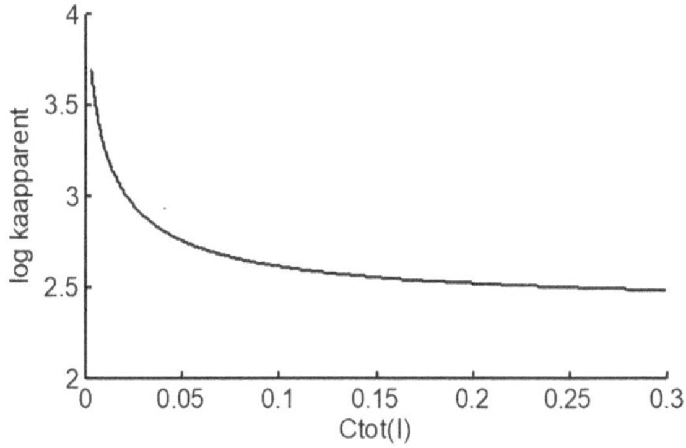

Fig. 4.4 Plot of log of apparent constant against different concentrations of entered component I, for the titration of 10 mL $C_A = 1 \times 10^{-4}$ M and I, with 0.14 mL $C_B = 1 \times 10^{-2}$ M, when C_I changes from 3.5×10^{-3} M to 3×10^{-1} M.

tunable range as a function of the total concentration of interferent I. This tuned range of apparent constant for any equilibrium reaction can be determined from the complete model of the related system. This is clear from Fig. 4.4 that under the mentioned condition, the apparent constant of the system could be tuned from $10^{3.69}$ to $10^{2.48}$.

While for many practical applications, it is not crucially important to have a constant value for $K'_{C'}$, or to maintain constant free [I], it is instructive to investigate the change in $K'_{C'}$ during a typical titration.

The change in [I] also depends on the initial concentrations of all involved components A, B, and I. A complete picture can show the effect of all controllable parameters on the tunable range of apparent constant. Fig. 4.5 shows the effect of different total concentrations of reactants (C_A and C_B) for the mentioned equilibrium system with the determined equilibrium constants. Twenty-eight samples with different C_A (10^{-6} to 1 M) and C_B (10^{-6} to 1 M) were simulated. Each surface in Fig. 4.5 displays the effect of changing the C_I for each sample with determining C_A and C_B on the tuned range of apparent constant. Indeed, for the equilibrium system of Eq. (4.2) with defined constants (K_C, $K_{A.I}$, and $K_{C.I}$), the tuned range of apparent constant is determined in all permissible ranges of total concentrations of all involved components (Fig. 4.5). Thus, the apparent constant of this system can be tuned from $10^{5.9}$ to $10^{2.48}$.

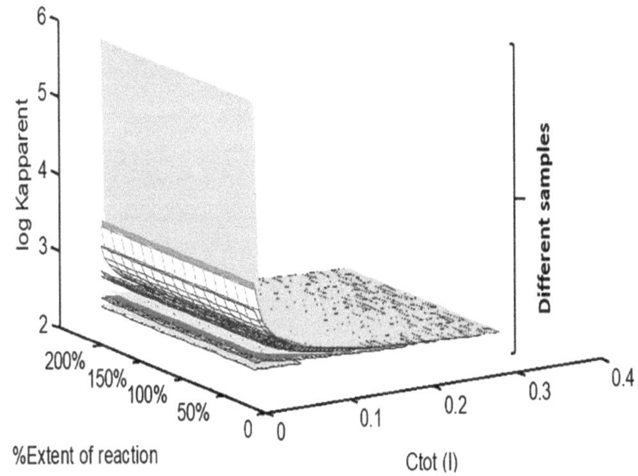

Fig. 4.5 Plot for the impact of total concentration of C_A, C_B, and C_I on the tuned range of apparent constant.

It can be generalized that for any equilibrium system, the tunable range of apparent constant is achieved in permissible ranges of all externally controllable parameters. This algorithm is generally applicable and solves chemical equilibrium problems of any degree of complexity. It calculates the species concentrations for a solution for which the total component concentrations, the chemical model, and its formation constants are known.

3. The application of proposed strategy in practice

There are several examples in the chemistry literature where the apparent equilibrium constants have been adjusted by variable such as solvent, micelle concentration, or some interfering chemical species (Adam, Bravo, & Mancini, 2014). One application of micelles in analytical chemistry is to study the pH indicators in order to shift the chemical equilibrium and therefore the pKa (Caselli, Mangone, Paolillo, & Traini, 2002; Love, Habarta, & Dorsey, 1984; Rostami, Karimvand, & Abdollahi, 2019). Garcia and Medel studied the mechanism of dye-surfactant interactions in micelles solutions in detail in Garcia and Sanz-Medel (1986). Safavi and Abdollahi (2001) have observed a significant shift in equilibrium and kinetic constants in micelle solutions. The effect of ionic liquids on the apparent acidity constants of acid-base equilibria of indicators was reported by several works (Adam et al., 2014; Kidani & Imura, 2010; Safavi, Abdollahi, Maleki, & Zeinali, 2009). Also, it has been shown that the investigation of some chemical equilibria has been possible by adjusting the apparent equilibrium constants under certain chemical-physical conditions (Okada, 1995; Woolley, Tomkins, & Hepler, 1972). Herein, the proposed strategy is applied for tuning the apparent equilibrium constants using other controllable variables, which can have many applications in designing efficient sensors for the determination and quantification of analytes. It is clear that the ability for tuning the equilibrium constants can be effective in the application of many equilibrium-based chemical processes.

Fundamentally, a chemical sensor is defined as device that has selective response to a particular analyte through a physical or chemical interaction and can be used for the qualitative or quantitative determination of the analyte. For pH determination, the device can be a pH meter or chemical compounds such as pH indicators in a solution (Hulanicki, Glab, & Ingman, 1991; Lim, Musto, Park, Zhong, & Suslick, 2008; Wencel, Abel, & McDonagh, 2014).

The efficiency of the most biosensors and chemo-sensors is governed by the amount of equilibrium constants, which define the sensitivity and selectivity of sensing processes. The difference in the affinity constant of different analytes with receptor makes their quantitative and qualitative analysis possible in competitive reactions. Usually, for improving the sensitivity or selectivity of sensors, the equilibrium constant has been tuned by optimizing the temperature of the reactions. Indeed, by tuning the equilibrium constant, the affinity of analytes to the receptors can be controlled and placed under optimum conditions.

Since the pH indicators are versatile for the detection of the pH of a solution, a great number of sensors utilize indicators. The color change of pH indicators is dependent on their protonation constant (logK). Thus, by tuning the apparent constants of indicators in different chemical environmental conditions, the apparent constant of pH indicators can be shifted to the desired value and thus also the transition pH. It means that the range of transition interval of an indicator can be tuned.

There are several reports in which pH indicators have been used as the probe in designed sensor arrays to the identification of different targets (Mac Craith et al., 1997; Musto, Lim, & Suslick, 2009; Sanchez-Barragan, Costa-Fernandez, & Sanz-Medel, 2005). In such sensors, the identification of an analyte is dependent on the pH of the chemical environment. Interaction of different targets by the appropriate receptors changes the solution of pH, in the transition interval of the probe indicator, which produces the signal for the identification of the target analytes. Since the detection method in such sensors is dependent on the changing pH, by tuning the apparent constant of pH indicators, efficient identification of a range of target analytes will be possible.

The following includes two case studies where the above idea is developed for tuning the apparent model and the design of parameters of the optimized pH indicators. The first system is simulated using equilibrium constants from the published chemical system. All simulations were performed in MATLAB using the standard Newton-Raphson algorithm (Maeder & Neuhold, 2007). This algorithm is generally applicable and solves chemical equilibrium problems of any degree of complexity. It calculates the species concentrations for a solution for which the total component concentrations, the chemical model, and its formation constants are known. In the real experimental data set for the second case study, the effect of micellar solution on apparent dissociation constants of three pH indicators (phenol red, neutral red, and methyl orange) was investigated, and chemical modeling provided a guideline for tuning of apparent protonation constants of pH indicators and consequently designing pH sensor in a broad range.

3.1 Case study 1

According to the IUPAC definition, the transition interval of an acid-base (pH) indicator is the concentration range of a chemical species that can be easily detected using the change of color of two forms of an indicator. The capability of indicators in changing their color near the equivalent point in acid-base titrations is dependent on their acidity dissociation constant (pKa). The conditional dependence of the apparent constant of some pH indicators has been reported previously (Safavi et al., 2009). It has been shown that the transition point and transition interval as well as the apparent constant of pH indicators have been altered under the influence of the different types of ionic liquids and even the concentrations of them. Also, the effect of different concentrations of micellar pseudo-phase on the apparent acidity constant of a pH indicator has been investigated (Babamoradi & Abdollahi, 2015; Sajedi-Amin, Abdollahi, & Naseri, 2021). Inspiring the strategy of tuning the apparent constants, the goal of this case study is to tune the pKa of acid-base indicator neutral red (NR) in micellar pseudo-phase for improving the range of its transition points and transition interval for more applications.

Aqueous micellar solutions have several applications in analytical chemistry; they can increase the solubility of solutes, and they can also shift the observed equilibrium constants, in particular the protonation constants of weak acids, e.g., pH indicators. Further, shifts in absorption spectra and changing apparent equilibrium and kinetic constants have been observed due to change in the chemical/physical properties of the solutions. These effects are due to interactions of solutes with micelles, i.e., the solutes can either remain in the aqueous phase or be dissolved in the micellar phase. Naturally, these different solvents will affect many properties of dissolved chemicals (Beltran, Codony, Granados, Izquierdo, & Prat, 1993; Beltran, Prat, & Codony, 1995; Dutta & Bhat, 1993; Funasaki, 1979; Underwood, 1982).

A complete model that describes the behavior of aqueous micellar systems is the partition model in which the solutes are distributed between the aqueous phase and the micellar pseudo-phase (Blatt, Chatelier, & Sawyer, 1984). The acid-base equilibrium of neutral red (NR) in aqueous micellar solutions of Brij-35 (polyoxyethylene (23) lauryl ether) is considered in the present case (Babamoradi & Abdollahi, 2015). Figure 4.6 displays the partition model for distributions of NR in aqueous micellar solution.

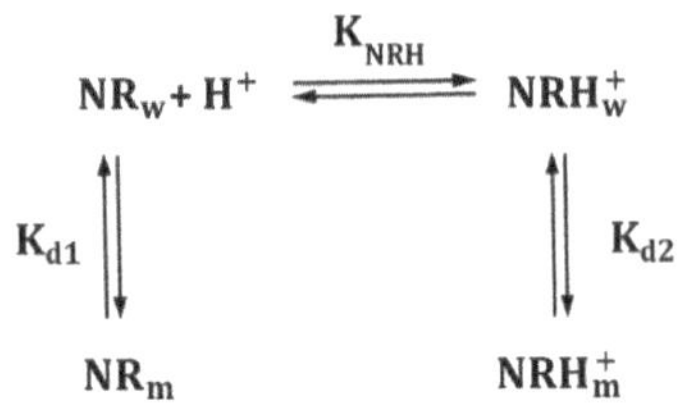

Fig. 4.6 The partition model for acid-base equilibrium of neutral red (NR) in an aqueous micellar solution of Brij-35.

Based on the above model, both forms of NR are distributed between the aqueous phase and micellar pseudo-phase. The corresponding distribution coefficients, K_{d1} and K_{d2}, are defined as follows:

$$K_{d1} = \frac{[NR]m}{[NR]w} \tag{4.9}$$

$$K_{d2} = \frac{[NRH+]m}{[NRH+]w} \tag{4.10}$$

where $[NRH^+]_w$ and $[NRH^+]_m$ are the local concentrations of NRH^+ in aqueous and micellar phases, respectively, and $[NR]_w$ and $[NR]_m$ represent the local concentrations of deprotonated form of NR in both phases, respectively.

By ignoring the distribution of protonation and deprotonation forms of acid between aqueous and micellar phases in the complete model, the simple apparent model of system is the same as Eq. (4.11):

$$H^+ + NR' \rightleftharpoons NRH^{+\prime} \quad K'_{NRH'} = \frac{[NRH^{+\prime}]}{[H^+][NR']} \tag{4.11}$$

We define $NRH^{+\prime}$ as the sum of the species NRH in both aqueous and micellar phases ($NRH+' = [NRH_w^+] + [NRH_m^+]$) and NR' as the sum of the species NR in both phases ($NR' = [NR_w] + [NR_m]$). Thus, as the total concentrations of species NRH'^+ and NR in different concentration of surfactant are constant, we can consider:

$$NRH'^+ = [NRH'^+]_w\,(1 - R) + [NRH'^+]_m\,(R) \tag{4.12}$$

$$NR' = [NR]_w\,(1 - R) + [NR]_m(R) \tag{4.13}$$

In these equations, R is the ratio of micellar volume (V_m) to the total volume of solution (V), which is defined as:

$$R = \frac{V_m}{V} = (C_m - cmc)\,V_\Phi \tag{4.14}$$

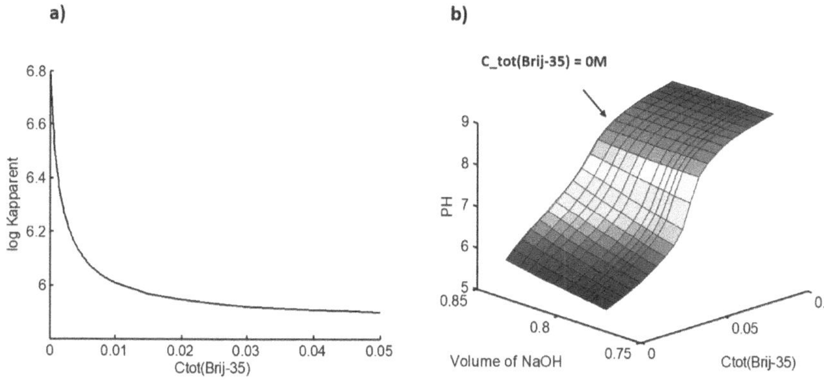

Fig. 4.7 (A) Plot of apparent K as a function of different concentrations of Brij-35 micelle. (B) Potentiometric titration curves for passive concentration range of Brij-35 in which the apparent constant is established.

where C_m is the total concentration of surfactant, cmc is the critical micelle concentration (cmc Brij-35 = 0.09 mM), and VΦ is the partial molar volume of micellized surfactant.

Then, by replacing Eqs. (4.12) and (4.13) in Eq. (4.11) and applying Eqs. (4.9) and (4.10), the relationship between apparent constant ($K'_{NRH'}$) and the constant of complete model can be extracted as:

$$K'_{NRH'} = K_{NRH}\frac{1 - R(1 + K_{d2})}{1 - R(1 + K_{d1})} \tag{4.15}$$

As it is expected, the apparent constant ($K_{NRH'}{}'$) can be controlled by the surfactant concentration. Of course, different surfactants have different distribution constants (K_{d1} and K_{d2}) as well as different cmc values.

In what follows, the effect of different total concentration of Brij-35 on the apparent acidity constant of neutral red (NR) is investigated, and the results are shown in Fig. 4.7. The permissible range of total concentration of Brij-35 in which the apparent constant is correct is shown in Fig. 4.7. Also, the potentiometric titration curves for NR in the absence and presence of micelle for the permissible concentration range of Brij-35 are displayed in Fig. 4.7. It is notable that the tuned range of apparent constant of NR in the presence of Brij-35 is different for another kind of micelle.

3.2 Case study 2: Analysis of experimental data (Rasouli, Maeder, & Abdollahi, 2021)

pH indicators are commonly used for pH determination, because of their availability and stability (Lin & Liu, 2000; Martinez-Olmos et al., 2011;

Mosley, Husheer, & Hunter, 2004). The disadvantage of pH indicators is their narrow dynamic range, and the linear range of pH indicator is limited over two pH units. There are two different methods, to overcome the narrow dynamic range and use pH indicators in a broader range, efficiently. An option is using a mixture of indicators in a solution (Bai et al., 2013; Barati, Shamsipur, & Abdollahi, 2016; Lin & Liu, 2000; Martinez-Olmos et al., 2011; Qi et al., 2015; Safavi & Bagheri, 2003; Santos et al., 2014). Another alternative is to use an array of pH indicators (Capel-Cuevas, Cuéllar, de Orbe-Payá, Pegalajar, & Capitán-Vallvey, 2011; Lin & Liu, 2000; Martinez-Olmos et al., 2011; Mosley et al., 2004) that each indicator is individually a sensor element and overall responses from all sensor elements are utilized for pH determination.

Herein, two aspects were considered in pH sensors optimization: (1) tuning equilibrium constants of indicators to achieve continuous broad pH range and (2) designing optimum combination (array or mixture) of pH indicators based on enriching analytical sensitivity.

In order to show the applicability of the idea for experimental data, three different indicators are selected (methyl orange, neutral red, and phenol red). pH is not covered by these three indicators efficiently, in the range between 0.5 and 10. In this study, as the chemical behavior of pH indicators in aqueous solution is well known, the chemical sensor was designed in aqueous solution. Also, the chemical models of pH indicator behavior in the micellar solution have been investigated.

Fig. 4.8 displays the range of pH, which can be determined by selected indicators; two different pH ranges can be covered by phenol red: 0–2 and 6.8–8.8 (red (dark gray in the printed version) solid lines); pH range of methyl orange is about 2.5–4.5 (yellow (light gray in the printed version) solid line), and neutral red covers pH from 5.8–7.8 (green (gray in the printed version) solid line). The dashed lines show the determinable pH range using indicators in the micellar solution with 0.05 M surfactant concentration, and indicators in micellar solution and aqueous solution together can cover pH range from 0.5 to 10 properly. One solution, for covering the

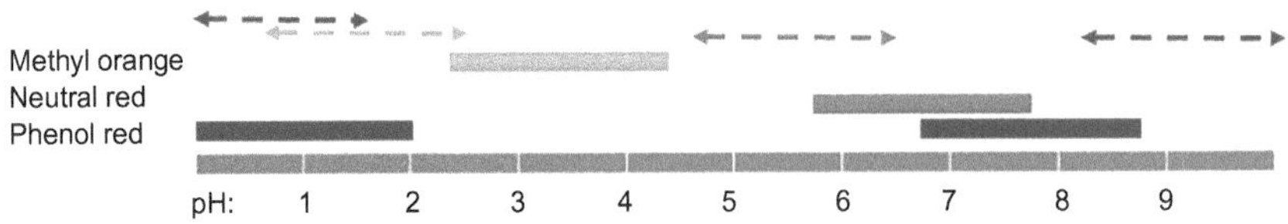

Fig. 4.8 pH range is covered by each indicator, and the *dashed lines* are the pH ranges covered by tuning pKa values of indicators.

complete pH range, is adding other indicators, with appropriate pKa values, to the existing set. In our case, at least one indicator with a pKa value of about 5.5 is required. Another option is tuning of pKa values of available indicators (methyl orange, neutral red, and phenol red) by adding the surfactant to the indicator solution. Distribution of indicators between micellar and aqueous phase changes their transition interval and transition point and consequently their apparent dissociation constants. The shift in apparent dissociation constant depends on the surfactant concentration and distribution constants of different forms of indicator in aqueous and micellar phases.

Eq. (4.15) states the relationship between apparent protonation constant and micelle concentration, and it can be seen, visually, in Fig. 4.9 for the selected indicators. The apparent constant shift for neutral red, methyl orange, and second protonation constant of phenol red is similar, and the pKa value decreases with increasing Brij-35 concentration. This behavior was predictable according to distribution constant values in Table 4.1 because the interaction between

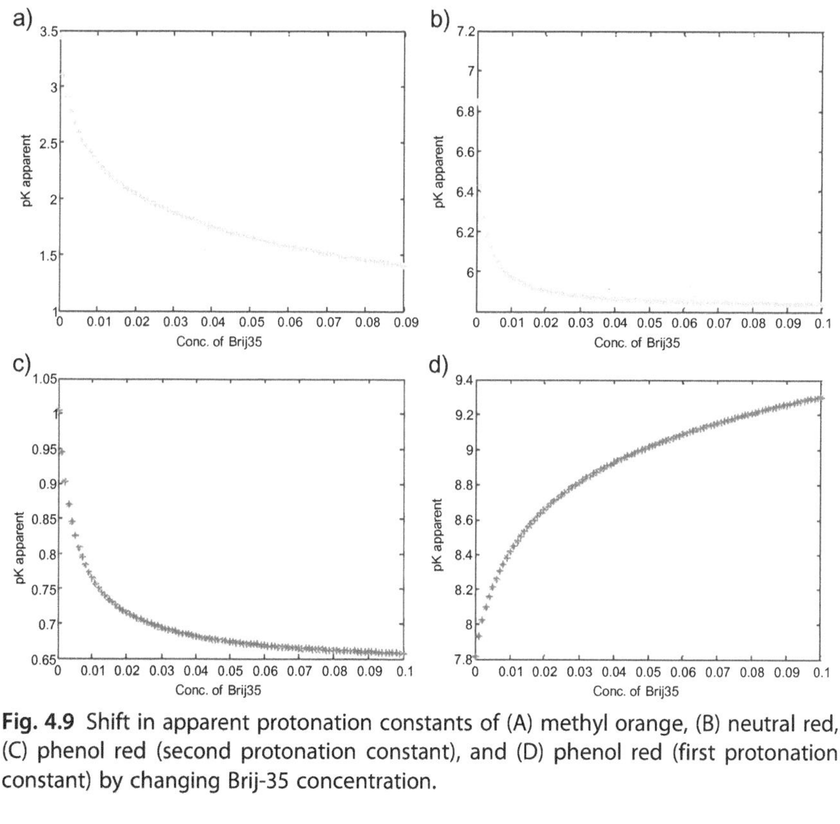

Fig. 4.9 Shift in apparent protonation constants of (A) methyl orange, (B) neutral red, (C) phenol red (second protonation constant), and (D) phenol red (first protonation constant) by changing Brij-35 concentration.

Table 4.1 Protonation constants and distribution coefficient of different forms of indicators in micellar aqueous solution.

	pK_a	K_{d1}	K_{d2}	K_{d3}
Phenol red	$pK_{a1} = 0.997 \pm 0.001$ $pK_{a2} = 7.830 \pm 0.007$	126.3 ± 0.5	289 ± 3	No observable interaction
Methyl orange	3.410 ± 0.009	No observable interaction	1103 ± 3	–
Neutral red	6.782 ± 0.004	200.6 ± 0.8	1827 ± 9	–

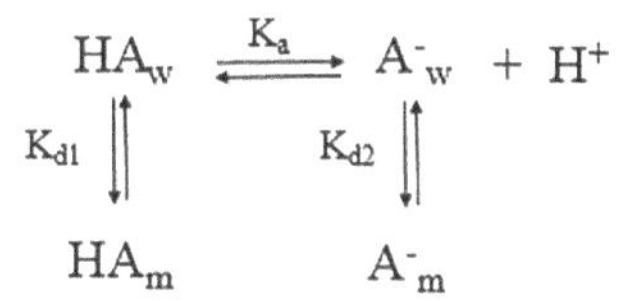

Fig. 4.10 Chemical model of acid-base equilibrium in aqueous micellar solutions.

deprotonated forms of acid with micelle proceeds the equilibrium condition and makes the acid stronger. However, the distribution of the protonated form to the micelle makes the acid weaker. So, the tuning of pKa values is possible by changing the surfactant concentration for each indicator. The chemical behavior of pH indicators in the micellar aqueous solution can be described based on the partition model, in which solute is distributed between aqueous and micellar pseudo-phase (Fig. 4.10). The distribution constants and spectral profiles of different forms of indicators were obtained using the nonlinear fitting algorithm. The chemical model and spectral profiles are required to predict optimum design for pH sensors.

There are two possible designs for three indicators: one design can be an array of the mixture of three indicators in aqueous solution and mixture in micellar pseudo-phase, while another design is an array of six sensor elements, i.e., three aqueous solutions and three micellar solutions. Fig. 4.11 depicts the two different designs of indicators, and the analysis results are summarized in Table 4.2.

The analytical sensitivity for the array of two mixtures is 1.36; this value for the array of six sensor elements is 0.62, while the sensitivity value is higher for this design, because it depends on the number of the wavelengths and it is not a suitable criterion for the comparison purpose. According to the analytical sensitivity values, the mixture of indicators has higher analytical sensitivity in comparison with the array. So based on this information, the array of mixtures in the aqueous and micellar phase is an optimum sensor.

The experiments were conducted based on the results of optimization using chemical modeling. Two batches of 39 solutions with pH range 0.5–10 were prepared, and solutions containing a mixture of three indicators one in aqueous solution and one in micellar solution were added to each group, respectively. In other words, for each pH, there are two samples in which aqueous solution of indicators is added to one sample and the micellar solution of indicators is added to another sample. Then, the spectra of each sample were recorded and augmented together, the PLS calibration model was constructed using 30 samples, and 9 samples were predicted using

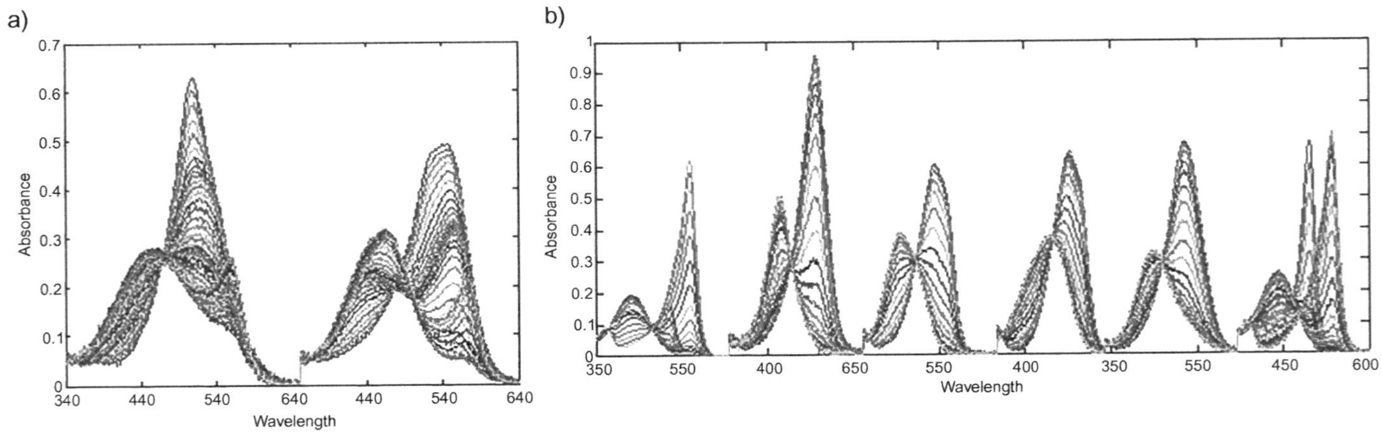

Fig. 4.11 (A) augmented mixture of three indicators in aqueous and micellar phases and (B) array of six sensor elements of indicators in micellar and aqueous solutions.

Table 4.2 Comparison of generalized analytical sensitivity of simulated single sensor elements (methyl orange, neutral red, and phenol red), mixture, and array.

	pH range	Sensitivity	Analytical sensitivity
Methyl orange: $D_{21\times252}$	2.41–4.41	0.98	4.67
Neutral red: $D_{21\times311}$	5.8–7.8	0.95	4.03
Phenol red: $D_{21\times201}$ $D_{21\times261}$	0–2	0.97	3.97
	6.8–8.8	1.05	4.70
Array of two mixtures: $D_{21\times311}$	0–10	0.50	1.36
Array of six sensor elements: $D_{21\times622}$	0–10	1.16	0.62

the calibration model. In the PLS model, the number of latent variables was selected based on the values of root mean square error prediction of leave-one-out cross-validation procedure, so the calibration model was constructed with five latent variables.

The optimum concentration of surfactant was picked out using the chemical modeling (information in Figs. 4.8 and 4.9), indicator with a pKa value of around 5.5 is required, and Fig. 4.9 shows that adding 0.05 M of surfactant to the solution of neutral red provides an indicator with a pKa of about 5.8. Also, adding 0.05 M of surfactant to methyl orange and phenol red can cover pH between 2 and 3.5 and 8.8 and 10, respectively.

The results of calibration for the sensor array of mixtures and each mixture (sensor elements), individually, are presented in Fig. 4.12. The results of PLS analysis confirm the assertion that a mixture of indicators in the aqueous solution is not sufficient for pH prediction by the PLS method. Fig. 4.12C and D demonstrates the result of experimental data analysis, for the mixture of three indicators in micellar and aqueous solutions, respectively. The predicted values of pH in the aqueous solution versus the measured pH by pH electrode are not in good agreement, particularly the pH values around 5. The obtained pH values, around 3 to 5, for the mixture of indicators in micellar solution (Fig. 4.12C) were not accurate, either. None of the sensor elements show proper efficiency for pH determination, individually. While the array of mixtures in micellar and aqueous solutions presents better analytical results, Fig. 4.12B displays the predicated pH using the PLS regression model versus measured pH, and the root mean square error of pH prediction is 0.1 for the prediction data set.

Previously, chemical model fitting was employed for quantitative pH determination using the mixture of these three indicators in aqueous

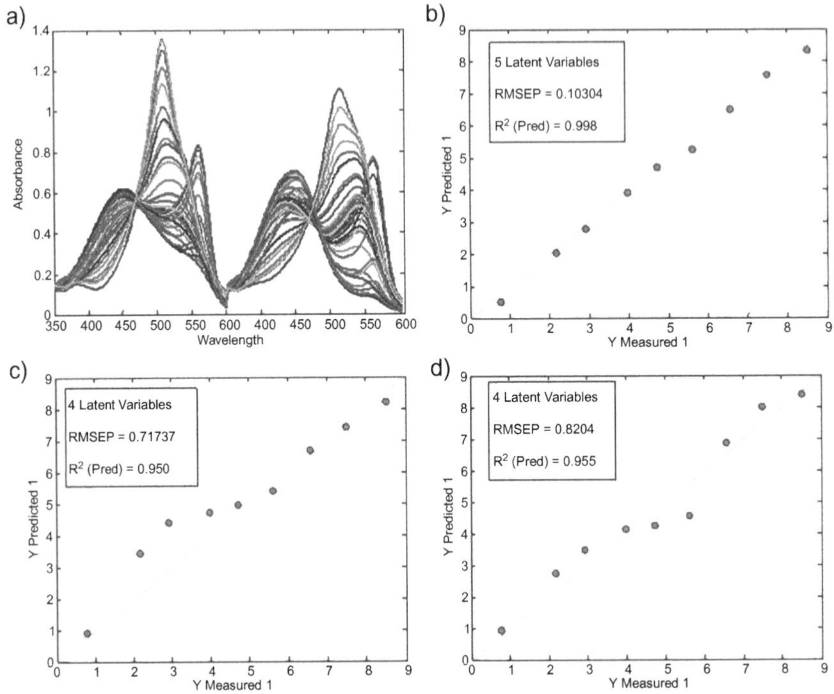

Fig. 4.12 (A) The experimental data of calibration set, which is an array of indicators mixture in aqueous and micellar phases, (B) the result of prediction of test samples pH using an array of mixtures in micellar and aqueous solutions, (C) the result of prediction test samples pH using a mixture of indicators only in micellar solution, and (D) the result of prediction test samples pH using a mixture of indicators in aqueous solution.

solution. The chemical model-based nonlinear algorithm was proposed, and for the analysis, the spectra of different forms of indicators were required as initial information (Rasouli, Abdollahi, & Maeder, 2020). Herein, the main point is designing sensors based on enhancement in the sensitivity (as an analytical figure of merit). The general analytical sensitivity was applied to apprehend optimum design of pH indicators in solution (a mixture or an array of pH indicators solutions). This idea of using parameters such as sensitivity can be applied for designing various chemical sensors where different sensing elements are used.

Like other optical sensors based on absorption, the pH indicators cannot be used in colored solution and another disadvantage is that the interaction between surfactant and indicators can be changed in the complicated matrix or mixed solvents. It is worth mentioning that we studied the behavior of hydrophilic pH indicators in aqueous solution and in the presence of micelle, and the pKa values of indicators were simply determined with titration in aqueous solution. However, different types of hydrophobic pH-sensitive probes have been introduced (Steinegger, Wolfbeis, & Borisov, 2020); in this case, a methodology was proposed to determine the pKa values of hydrophobic colorimetric pH-sensitive probes, and also, it is applicable in any water-immiscible nanomaterials (Xie, Zhai, Jarolímová, & Bakker, 2016).

References

Adam, C., Bravo, M. V., & Mancini, P. M. (2014). Molecular solvent effect on the acidity constant of protic ionic liquids. *Tetrahedron Letters*, *55*(1), 148–150.

Babamoradi, H., & Abdollahi, H. (2015). Study of monoprotic acid–base equilibria in aqueous micellar solutions of nonionic surfactants using spectrophotometry and chemometrics. *Spectrochimica Acta Part A: Molecular and Biomolecular Spectroscopy*, *149*, 851–859.

Bai, Z., Chen, R., Si, P., Huang, Y., Sun, H., & Kim, D.-H. (2013). Fluorescent pH sensor based on ag@ SiO2 core–shell nanoparticle. *ACS Applied Materials & Interfaces*, *5*(12), 5856–5860.

Barati, A., Shamsipur, M., & Abdollahi, H. (2016). Carbon dots with strong excitation-dependent fluorescence changes towards pH. Application as nanosensors for a broad range of pH. *Analytica Chimica Acta*, *931*, 25–33.

Beltran, J., Codony, R., Granados, M., Izquierdo, A., & Prat, M. (1993). Acid—base and distribution equilibria of 5, 7-dichloro-2-methyl-8-hydroxyquinoline in Brij-35 micellar media solutions. *Talanta*, *40*(2), 157–165.

Beltran, J., Prat, M., & Codony, R. (1995). Interaction of 5, 7-dichloro-2-methyl-8-hydroxyquinoline with ionic micelles. *Talanta*, *42*(12), 1989–1997.

Blatt, E., Chatelier, R. C., & Sawyer, W. H. (1984). Partition and binding constants in micelles and vesicles from fluorescence quenching data. *Chemical Physics Letters*, *108*(4), 397–400.

Capel-Cuevas, S., Cuéllar, M., de Orbe-Payá, I., Pegalajar, M., & Capitán-Vallvey, L. (2011). Full-range optical pH sensor array based on neural networks. *Microchemical Journal*, *97*(2), 225–233.

Caselli, M., Mangone, A., Paolillo, P., & Traini, A. (2002). Determination of the acid dissociation constant of bromocresol green and cresol red in water/AOT/isooctane reverse micelles by multiple linear regression and extended principal component analysis. *Annali di Chimica*, *92*(5–6), 501–512.

Dutta, R. K., & Bhat, S. N. (1993). Association of neutral red with micelles and its effect on the p K a. *Canadian Journal of Chemistry*, *71*(11), 1785–1791.

Funasaki, N. (1979). Micellar effects on the kinetics and equilibrium of chemical reactions in salt solutions. *Journal of Physical Chemistry*, *83*(15), 1998–2003.

Garcia, M. D., & Sanz-Medel, A. (1986). Dye-surfactant interactions: A review. *Talanta*, *33*(3), 255–264.

Hulanicki, A., Glab, S., & Ingman, F. (1991). Chemical sensors: Definitions and classification. *Pure and Applied Chemistry*, *63*(9), 1247–1250.

Karimvand, S. K., Maeder, M., & Abdollahi, H. (2019). A chemical equilibrium modelling strategy for tuning the apparent equilibrium constants of the chemical systems. *Analytica Chimica Acta*, *1049*, 29–37.

Kidani, K., & Imura, H. (2010). Solvent effect of ionic liquids on the distribution constant of 2-thenoyltrifluoroacetone and its nickel (II) and copper (II) chelates and the evaluation of the solvent properties based on the regular solution theory. *Talanta, 83*(2), 299–304.

Lim, S. H., Musto, C. J., Park, E., Zhong, W., & Suslick, K. S. (2008). A colorimetric sensor array for detection and identification of sugars. *Organic Letters, 10*(20), 4405.

Lin, J., & Liu, D. (2000). An optical pH sensor with a linear response over a broad range. *Analytica Chimica Acta, 408*(1–2), 49–55.

Love, L. C., Habarta, J. G., & Dorsey, J. G. (1984). The micelle-analytical chemistry interface. *Analytical Chemistry, 56*(11), 1132A–1148A.

Mac Craith, B., Mc Donagh, C., McEvoy, A., Butler, T., O'keeffe, G., & Murphy, V. (1997). Optical chemical sensors based on sol-gel materials: Recent advances and critical issues. *Journal of Sol-Gel Science and Technology, 8*(1), 1053–1061.

Maeder, M., & Neuhold, Y.-M. (2007). *Practical data analysis in chemistry*. Elsevier.

Martinez-Olmos, A., Capel-Cuevas, S., López-Ruiz, N., Palma, A., De Orbe, I., & Capitán-Vallvey, L. (2011). Sensor array-based optical portable instrument for determination of pH. *Sensors and Actuators B: Chemical, 156*(2), 840–848.

Mosley, L. M., Husheer, S. L., & Hunter, K. A. (2004). Spectrophotometric pH measurement in estuaries using thymol blue and m-cresol purple. *Marine Chemistry, 91*(1–4), 175–186.

Musto, C. J., Lim, S. H., & Suslick, K. S. (2009). Colorimetric detection and identification of natural and artificial sweeteners. *Analytical Chemistry, 81*(15), 6526–6533.

Okada, T. (1995). Non-aqueous capillary electrophoretic separation of polyethers and evaluation of weak complex formation. *Journal of Chromatography A, 695*(2), 309–317.

Qi, J., Liu, D., Liu, X., Guan, S., Shi, F., Chang, H., et al. (2015). Fluorescent pH sensors for broad-range pH measurement based on a single fluorophore. *Analytical Chemistry, 87*(12), 5897–5904.

Rasouli, Z., Abdollahi, H., & Maeder, M. (2020). Generalized indicator-based determination of solution pH. *Analytica Chimica Acta, 1109*, 90–97.

Rasouli, Z., Maeder, M., & Abdollahi, H. (2021). Using chemical modeling for designing of optimal pH sensor based on analytical sensitivity enhancement. *Microchemical Journal, 168*, 106450.

Ringbom, A. (1963). *Complexation in Analytical Chemistry, Interscience, New York.*

Rostami, A., Karimvand, S. K., & Abdollahi, H. (2019). Model-based description of indicator displacement assay sensor arrays for quantitation of mixtures. *Journal of Chemometrics, 33*(11), e3186.

Safavi, A., & Abdollahi, H. (2001). Spectrophotometric study on micelle-mediated shift in kinetic and equilibrium of complex formation between Ni2+ and 2-amino-cyclopentene-1-dithiocarboxylic acid. *Microchemical Journal, 69*(2), 69–77.

Safavi, A., Abdollahi, H., Maleki, N., & Zeinali, S. (2009). Investigation of the role of ionic liquids in tuning the pKa values of some anionic indicators. *Journal of Solution Chemistry, 38*(6), 753–761.

Safavi, A., & Bagheri, M. (2003). Novel optical pH sensor for high and low pH values. *Sensors and Actuators B: Chemical, 90*(1–3), 143–150.

Sajedi-Amin, S., Abdollahi, H., & Naseri, A. (2021). The influence of coupled kinetic-equilibrium in breaking the correlation between kinetic parameters. *Chemometrics and Intelligent Laboratory Systems*, 104349.

Sanchez-Barragan, I., Costa-Fernandez, J., & Sanz-Medel, A. (2005). Tailoring the pH response range of fluorescent-based pH sensing phases by sol–gel surfactants co-immobilization. *Sensors and Actuators B: Chemical, 107*(1), 69–76.

Santos, L., Neto, J. P., Crespo, A., Nunes, D., Costa, N., Fonseca, I. M., et al. (2014). WO3 nanoparticle-based conformable pH sensor. *ACS Applied Materials & Interfaces, 6*(15), 12226–12234.

Steinegger, A., Wolfbeis, O. S., & Borisov, S. M. (2020). Optical sensing and imaging of pH values: Spectroscopies, materials, and applications. *Chemical Reviews, 120*(22), 12357–12489.
Tammelin, L., & Mogensen, S. (1952). Microquantitative determination of calcium as Murexide complex in presence of magnesium. *Acta Chemica Scandinavica, 6*.
Underwood, A. (1982). Dissociation of acids in aqueous micellar systems. *Analytica Chimica Acta, 140*(1), 89–97.
Wencel, D., Abel, T., & McDonagh, C. (2014). Optical chemical pH sensors. *Analytical Chemistry, 86*(1), 15–29.
Woolley, E. M., Tomkins, J., & Hepler, L. G. (1972). Ionization constants for very weak organic acids in aqueous solution and apparent ionization constants for water in aqueous organic mixtures. *Journal of Solution Chemistry, 1*(4), 341–351.
Xie, X., Zhai, J., Jarolímová, Z., & Bakker, E. (2016). Determination of pKa values of hydrophobic colorimetric pH sensitive probes in nanospheres. *Analytical Chemistry, 88*(6), 3015–3018.

CHAPTER 5

The analytical/measurement sources of multivariate errors. A case study: Detecting microplastics in sand

Reaha Goyetche[a], José Manuel Amigo[a,b], and Leire Kortazar[a,c]
[a]Department of Analytical Chemistry, University of the Basque Country, Leioa, Spain
[b]IKERBASQUE, Basque Foundation for Science, Bilbao, Spain
[c]Plentzia Marine Station, University of the Basque Country (PiE-UPV/EHU), Plentzia, Spain

1. Introduction

1.1 Adding "microplastics" to the dictionary

Though a world without plastics is scarcely imaginable today, wide-scale plastic production only began in 1950. In the intervening 70 years, the plastics industry manufactured some 8300 million metric tons (Mt) of virgin plastics—a full 60% of which is sitting in landfills or accumulating in the natural environment (Geyer, Jambeck, & Law, 2017). According to the 2020 Plastics Europe Report, global production of plastics reached 368 million tons—an increase of 9 million tons from the previous year. Packaging, which is generally single use and quickly discarded, represented a full 39.4% of the 50.7 tons of plastics used in Europe in 2019 (Plastics Europe, 2020). During the inaugural International Research Workshop on the Occurrence, Effects and Fate of Microplastic Marine Debris hosted by NOAA in 2008, microplastics were defined as all plastic particles smaller than 5 mm in size, based on their potential for ingestion by biota (Arthur & Baker, 2008). Later guidance further refined the definitions as listed in Table 5.1 (Guidance on Monitoring of Marine Litter in European Seas, 2013).

Theoretical definitions aside, microplastic size ranges studied often depend more on the limits imposed by sampling methods and instrumental sensitivity. From a practical standpoint, microplastics are often characterized in two fractions, those from 5 mm to 500 μm in size and those <500 μm, as 500 μm tends to be the lower limit at which particles can be manually sorted (Hidalgo-Ruz,

Machine Learning and Pattern Recognition Methods in Chemistry from Multivariate and Data Driven Modeling
https://doi.org/10.1016/B978-0-323-90408-7.00003-4

Table 5.1 Definitions of size ranges of plastics proposed by the MSFD technical subgroup on marine litter.

Plastic debris nomenclature	Size range
Macroplastic	>2.5 cm
Mesoplastic	5 mm–2.5 cm
Large microplastic	1 mm–5 mm
Small microplastic	1 μm–1 mm
Nanoplastic	<1 μm

Gutow, Thompson, & Thiel, 2012). To the end of monitoring and characterizing microplastics in sediments, some protocols recommend a lower size limit of 100 μm based on sampling capabilities (Frias et al., 2018). Microplastics are further characterized according to their origin. Primary microplastics are those manufactured to microscopic sizes, whereas secondary microplastics are derived from the breakdown of larger plastic particles in the environment (Frias & Nash, 2019). With the persistence of hundreds of thousands of years, all of the plastic ever introduced into the environment remains unmineralized; instead, plastics fragment is due to ultraviolet (UV) degradation, oxidation, and abrasion (Barnes, Galgani, Thompson, & Barlaz, 2009). This degradation means that methods of characterization that depend on spectral libraries can be expected to fall short, given that the spectra of weathered plastics differ greatly from those of standard plastics (Dong et al., 2020). Taking all of these into account, the very characteristics that make plastics useful—persistent, inexpensive, water-resistant, versatile—make them worrisome in marine and terrestrial environments.

1.2 Persistent, ubiquitous—And miscounted?

The persistence of plastics contributes to their ubiquity in the natural environment. Even the most seemingly pristine and far-flung sites are not safe from plastic pollution. Thanks to long-range transport, plastics can be found in places as remote as polar ecosystems (Halsband & Herzke, 2019). Their size, meanwhile, allows for uptake by living organisms. The recent (and disputed) discovery of microplastics in human placentas is one concerning example of this phenomenon (Ragusa et al., 2021).

There are ecological consequences to this proliferation of microplastics. In the marine environment, microplastics are ingested at trophic levels ranging from zooplankton to marine mammals, and including filter feeders such as *Mytilus edulis*. The plastics themselves are not the only cause for concern, as their large surface area invites the adsorption of persistent organic pollutants (POP) and persistent, bioaccumulative, and toxic (PBT) chemicals (Kershaw, 2015). Lacking quantitative methods means that little is known about the co-bioaccumulation of microplastics with adsorbed toxins, which are known to include polycyclic aromatic hydrocarbons and heavy metals (Xu, Ma, Ji, Pan, & Miao, 2020).

In the oceans, the problem extends to all five of the main ocean compartments—the surface, the water column, the seafloor, biota, and on the shoreline or buried in intertidal sediments (Kershaw & Rochman, 2015). Following 680 net pulls on 24 separate expeditions across the Atlantic, Pacific, and Indian Oceans, as well as the Mediterranean Sea, Eriksen et al. (2014) estimate that there are over 5 trillion plastic particles floating in the ocean. Concentrations as high as 890,000 pieces of plastic per square kilometer were measured in the Mediterranean Sea (Eriksen et al., 2014).

However, there is a gap in the understanding of the exchange between compartments and a lack of knowledge of the quantity of plastic in each of the compartments, except the surface waters. Though the concentrations of plastics observed at sea are concerning, 4–23 times more plastic is deposited on land than into oceans each year (Horton, Walton, Spurgeon, Lahive, & Svendsen, 2017). Beaching, or the deposition of plastics on beaches by tides and waves, is one process by which this occurs. In most cases, plastics with lower density than seawater float in surface waters and wash up on shores, especially at times of high wind speeds, wave height, and daily river flow in estuaries (Lefebvre et al., 2021). Even so, there is an enormous mismatch between the amount of plastic believed to enter the natural environment each year and the amounts observed by scientists, in what is called the "missing plastic problem" (Cressey, 2016). The previously mentioned sampling expeditions found an entire order of magnitude less microplastic than predicted in their pre-expedition estimations (Eriksen et al., 2014). As such, understanding terrestrial impacts and refining the techniques to identify and quantify plastic litter on land is of the utmost importance. Furthermore, this undercount of microplastic undermines the efforts to calculate the risk arising from plastics in the natural environment, because it is difficult to provide a confident measure of exposure levels (Kershaw, 2015).

1.3 Hyperspectral imaging as a method for microplastics detection

One major challenge in microplastics research is the lack of standardized analytical and quantification techniques, making it difficult to compare results across studies (Horton et al., 2017). Many of the common methodologies include density separation of plastics from solid matrices followed by filtration, which is time-consuming and can lead to the loss of material, especially when polymers denser than common separation solutions are present (Frias et al., 2018). There are several chemical analytical methods that rely on thermodegradation and visual analysis. Two common thermoanalytical methods are pyrolysis gas chromatography coupled with mass spectrometry and thermogravimetric analysis, but these will not be discussed as they are destructive and provide a measurement of percent weight rather than particle dimensions or amount of particles (Käppler et al., 2016). Visual analysis is a valid technique to identify plastic particles as small as 0.25–2 mm based on their physical features, but this method is highly subject to error and identification varies greatly between observers (Ruggero, Gori, & Lubello, 2020). Stereomicrosopy is capable of analyzing samples in the micron order of magnitude and providing physical characteristics and abundance; however, it does not provide information on composition (Yang, Zhang, Kang, Wang, & Wu, 2021).

Given the limitations of methodologies based on thermodegradation and visual analysis, hyperspectral imaging has emerged as a reliable, non-destructive, non-invasive technique to characterize microplastic waste, thereby expanding the possibilities for plastic pollution monitoring (Serranti, Palmieri, Bonifazi, & Cózar, 2018). For the purposes of characterizing and identifying microplastics, the spectra can be collected by Raman, FTIR, and NIR—although a wide variety of other techniques are possible. Hyperspectral imaging refers to an analytical technique in which images are collected in such a way that a spectrum in a full wavelength range is collected for each pixel. The spectral resolution, then, refers to the interval between each wavelength collected in the selected wavelength range, with a smaller interval corresponding to a higher spectral resolution (Amigo, 2020).

In this context, an image shall be defined as a two-dimensional representation of a surface. Images are collected by an instrument capable of obtaining X-Y information correlated with the surface in question. Subsequently, the term "pixel" must be defined. A pixel is the smallest individual spatial subdivision of an image. Following this definition, spatial resolution can be expressed as X × Y for a given image, where X is the number of pixels per row and Y is the number of pixels per column. However, when discussing the spatial resolution, it is important to remember that each pixel represents a given surface area within the image, as X and Y correspond to physical distances (Amigo, 2020). The visual representation of a hyperspectral image is a hypercube defined in three dimensions: the X and Y spatial directions and the spectral (λ) dimensions. As such, the mathematical notation to represent a hypercube D is X × Y × λ (Amigo, Babamoradi, & Elcoroaristizabal, 2015) (Fig. 5.1).

1.4 Chemometrics

Given the overwhelming amount of spectral and spatial information produced, advanced machine learning methods [a.k.a. chemometrics (Massart, 1997)] are required to glean meaningful results from the massive quantities of data produced by the instruments used in analytical chemistry. In the context of microplastic studies, classification strategies are the most used chemometric methodologies (Da Silva, Murphy, Amigo, Stedmon, & Strand, 2020).

1.5 Principal component analysis

Principal component analysis (PCA) is an unsupervised exploratory chemometric tool. That is to say, it explores variation within the multivariate datasets, but it is not a method of classification. Applied correctly, PCA can provide an overview of multivariate data, shed light on the interrelatedness of variables and samples, and quantify patterns (Bro & Smilde, 2014). More specifically, PCA attempts to decrease the dimensionality of complex data sets by finding orthogonal directions in the multivariate space capable of describing the maximum variance as:

$$D = TP^T + E \tag{5.1}$$

Using the hyperspectral image in Fig. 5.1 as an example, D is the unfolded matrix with dimensions of $XY \times \lambda$, containing the data for XY pixels and λ variables. T, the score vector or scores, has dimensions of $(XY \times R)$, and P, the loading vector or loadings, has dimensions of $(\lambda \times R)$, where R refers to the number of principal components (Da Silva et al., 2020). E is the residual vector (Bro & Smilde, 2014).

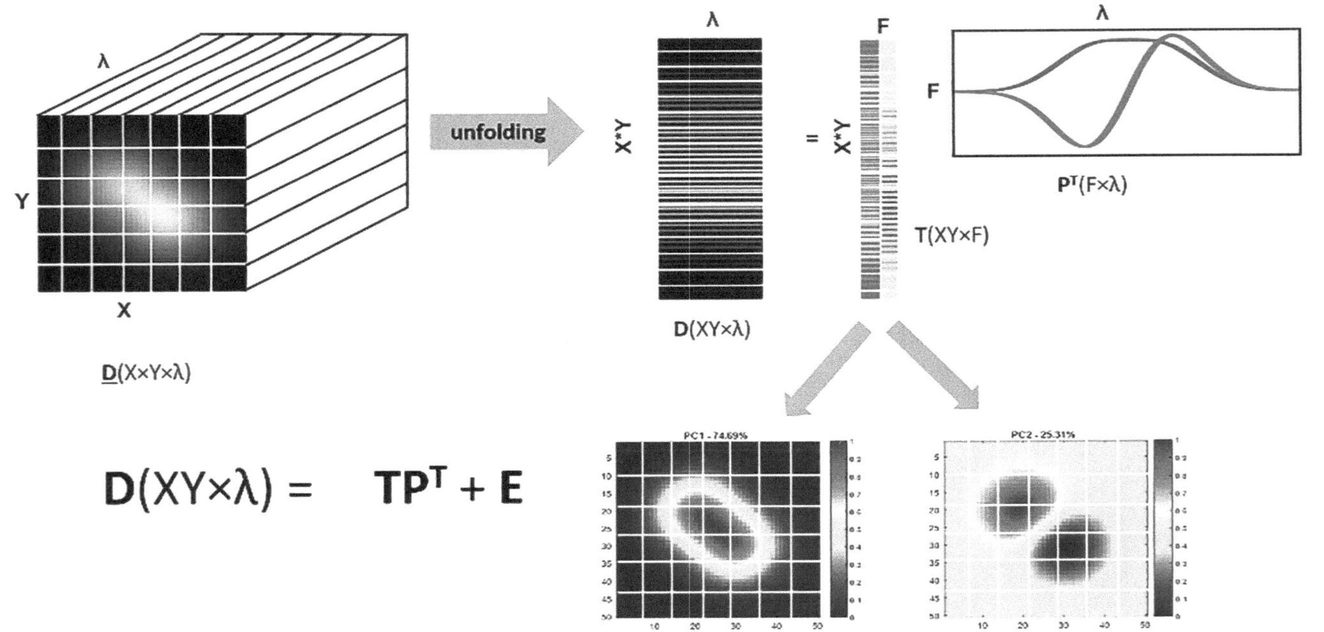

Fig. 5.1 A hypercube, D, must be unfolded into a matrix, which must be performed before applying modeling techniques: in this case, a principal component analysis model. *(Reproduced from Marini, F. and Amigo, J.M. (2019). Chapter 2.4 - Unsupervised exploration of hyperspectral and multispectral images. In: Data handling in science and technology, Vol. 32 (pp. 93–114). https://doi.org/10.1016/B978-0-444-63977-6.00006-7, with written permission of Elsevier.)*

1.6 Classification

In contrast with exploratory techniques such as PCA, classification refers to supervised multivariate analytical methods that are designed to sort samples into classes based on a series of measurements or theoretical values (Marini, 2010). These measurements or theoretical values define the classes, or groups, of objects, and can be referred to as training sets (Marini, 2010). A validation step (internal cross-validation or external validation) is necessary to ensure the reliability of a model (Amigo, 2020).

Two of the primary categories of classification techniques are pure classification (discriminant) techniques, wherein the focus is on differences between samples of different classes, and class-modeling techniques, wherein the focus is on the similarities among samples of the same class (Marini, 2010). This introduction will focus on linear classification methods, but it is worth commenting that non-linear classification methods (e.g., artificial neural networks and k-means clustering) are also being used to a certain extent.

1.6.1 SIMCA

Soft independent modeling by class analogy (SIMCA) is a classification method based on the use of principal component models to describe each of the classes (Marini, 2010). Each class in a training set is a quantitatively defined group of objects. The model for each class (e.g., Category 1) is then described by a set number, A, of principal components, according to:

$$D_1 = T_A P_A^T + E \tag{5.2}$$

where D_1 is the submatrix of the full data set containing only the samples of Category 1. T_A and P_A are the matrices composed of the first A scores and loading vectors, respectively, and E is the residuals (Marini, 2010).

1.6.2 PLS-DA

As its name implies, partial least-squares discriminant analysis (PLS-DA) is a discriminant classification technique based on partial least-squares (PLS) algorithm. In the case of hyperspectral imaging, PLS-DA treats each pixel as if it were a sample and then assigns it to a class based on the regression model developed from the training set. Each pixel is projected onto a regression model and the probability of the pixel corresponding to each class is calculated, and a threshold determines which class the pixel is assigned to

(Amigo et al., 2015). The sample is then assigned to the class it has the highest probability of belonging to (Marini, 2010).

1.6.3 Validation

The validation parameters used in a classification model are sensitivity, specificity, and error rate, all of which are generally expressed in percentages. These parameters are based on the number of true positives (TP), false positives (FP), true negatives (TN), and false negatives (FN) during the internal cross-validation or external validation step (Amigo et al., 2015).

$$\begin{aligned} \text{Sensitivity} &= \frac{TP}{(TP + FN)} \\ \text{Specificity} &= \frac{TN}{(FP + TN)} \\ \text{Error rate} &= 1 - \frac{(\text{Sensitivity} + \text{Specificity})}{2} \end{aligned} \tag{5.3}$$

Sensitivity refers to the ability of the model to confirm that a sample belongs to the class to which it has been assigned, and specificity is the ability of the model to confirm that a sample does not belong to the classes to which it has not been assigned (Amigo et al., 2015).

1.7 Objectives of this book chapter

Despite the abovementioned promising benefits of NIR-HSI, there are important analytical concerns that require a deep understanding of the interactions between NIR radiation, microplastics, and the matrices in which microplastics are located. These concerns beg a series of questions:

- Signal interference: Will the spectrum of the microplastic be the same as the spectrum of pure plastic? If so, to what extent can the signal of a microplastic be differentiated from the signals of the surrounding elements?
- Scattering: Will the spectrum of a pixel containing microplastic be affected by the scattering promoted by the surrounding elements? If so, to what extent will this effect change the microplastic spectrum relative to the spectrum of pure plastic?
- Spatial resolution: What is the smallest spatial level at which we can be confident that the microplastic signal will still be visible?
- Spectral resolution: Is there a combination of wavelengths in the spectra that remain selective for the microplastic?

This chapter gives some answers to the abovementioned questions and settles the basis for understanding the extent to which NIR-HSI and classification methods can be used to detect microplastics. The hypothesis is that NIR-HSI, together with the proper classification method, is a viable technique to identify microplastics and measure their size directly in the sand, avoiding time-consuming separation techniques. For this aim, we collect a NIR-HSI image of a very well-standardized microparticle of known identity (acrylonitrile butadiene styrene) placed in the well-standardized sand (acting as marine sediment) to assess the realism of classification chemometric techniques. This will help us to understand the extent to which the spectra of microplastics are influenced by their surroundings.

This chapter is the beginning of a series of research actions being developed at our research group. Therefore, the experimental setups and the results are just the starting point that sets the basis for future more chemometric-elaborated answers.

2. Materials and methods

2.1 Acrylonitrile butadiene styrene (ABS) pellet

The polymer used was the acrylonitrile butadiene styrene (ABS). It was in the form of pellets of approximately 5 mm in diameter. ABS was chosen from the list of polymers produced in the greatest quantities in Europe (Plastics Europe, 2020). Microparticles ranging in size from 100 μm to 2 mm were produced from the pellet using mechanical abrasion.

The identity and size of ABS were confirmed using Raman microscopy. An inVia confocal Raman spectrometer coupled to a DMLM Leica microscope (Renishaw, Gloucestershire, UK) was used to measure the size and confirm the identity of the polymer. A still image of each microparticle on a glass plate was collected with 20 × magnification. The still image collected of each individual microplastic particle was used to calculate the dimensions. A 785-nm excitation laser was used with 350 mW output power and 150 mW set as 100% laser power.

2.2 Sample preparation

Samples of sand were collected from the top surface layer (approximately 2 cm deep) of Plentzia Beach 1 h after high tide in Plentzia, Spain (43° 24′40″N 2°56′46″W). The sand was rinsed with tap water, and any floating

Fig. 5.2 Left, Petri dish of sand doped with one ABS microplastic, along with reference plastic pellets. Right, the hyperspectral camera used for this experiment.

debris was removed and dried in an oven at 100°C overnight. The sand was then placed in an oven at 450°C for 1 h to remove any remaining traces of organic material. The sand was stored in a glass container.

Spiked samples were prepared by filling Petri dishes with the sand to a depth of 0.5 cm and placing one microparticle on the surface of the sand (Fig. 5.2). Reference pellets of ABS were placed alongside the petri dish on the sample tray to provide reference spectra.

2.3 Near infrared hyperspectral imaging

Near infrared (NIR) hyperspectral images were collected using a Headwall Photonics NIR hyperspectral camera (Model 1002A-00371, Fitchburg, MA, USA). X-Control software was used to calibrate the camera settings, and Hyperspec Rev. A.1.13.1 software was used to set the acquisition parameters. Images were collected in the spectral range from 1000 to 1700 nm with a spectral resolution of 4.85 nm and 142 bands with an acquisition time of 100,000 μs, a temperature of 260 K, and a focal distance of 30 cm. The spatial resolution was set to 10 μm. The light source was aimed at the sample at a 45° angle. Calibration was performed by taking a 0% reflectance measurement with the objective closed and 100% reflectance measurement with a Spectralon plate. The camera measures in line-scan modality. Scan length was adjusted as necessary to record all of the samples (Fig. 5.2).

2.4 Data analysis and software

The hyperspectral images were analyzed in MatLab (MathWorks, Natick, MA, USA) using HYPER-Tools (Mobaraki & Amigo, 2018) to perform spectral preprocessing, PCA, SIMCA, and PLS-DA. The data were preprocessed by using two strategies: standard normal variate (SNV) and first derivative.

To create training sets for classification, 200 spectra from pure sand and 200 spectra from the reference ABS pellet were extracted to form a final spectral library of 400 spectra. The pure sand and reference ABS pellet were measured simultaneously with the samples to avoid problems with variability (light conditions, focal distance, or spatial accuracy of the camera). All models (SIMCA with SNV; SIMCA with first derivative; PLS-DA with SNV; and PLS-DA with first derivative) were validated using random-subsets cross-validation and external validation (using 75% of the spectra for training and cross-validation, and 25% of the spectra for external validation). The datasets for training and external validation were randomly split.

3. Results and discussion

3.1 Raman microscopy

Prior to measurement with NIR-HSI, the identity and the dimensions of the microplastic were confirmed via Raman microscopy (Fig. 5.3).

The area of the ABS microparticle was calculated to be 437,250 $\pm 6949\,\mu m^2$ with a major axis of $922 \pm 6\,\mu m$. The characteristic peaks present in the Raman spectra confirmed the identity of the microparticles to be ABS based on comparison with the Raman spectra of reference ABS pellets.

3.2 NIR-HSI spectra

Given that the microparticle of ABS was manually placed on top of the sand, its general location was approximately known and the region of interest (RoI) was selected to include this location and the surrounding sand as background (Fig. 5.4A), while the microparticle is located inside the red (dark gray in the printed version) circle.

The principal peaks for ABS occur at 1145, 1198, and 1412 nm, as seen in the average spectrum derived from the spectra of the reference ABS pellet (Fig. 5.4B and C). In contrast to the spectra of the reference ABS, there is more baseline shift and scattering in spectra corresponding to regions known to contain only sand (Fig. 5.4D).

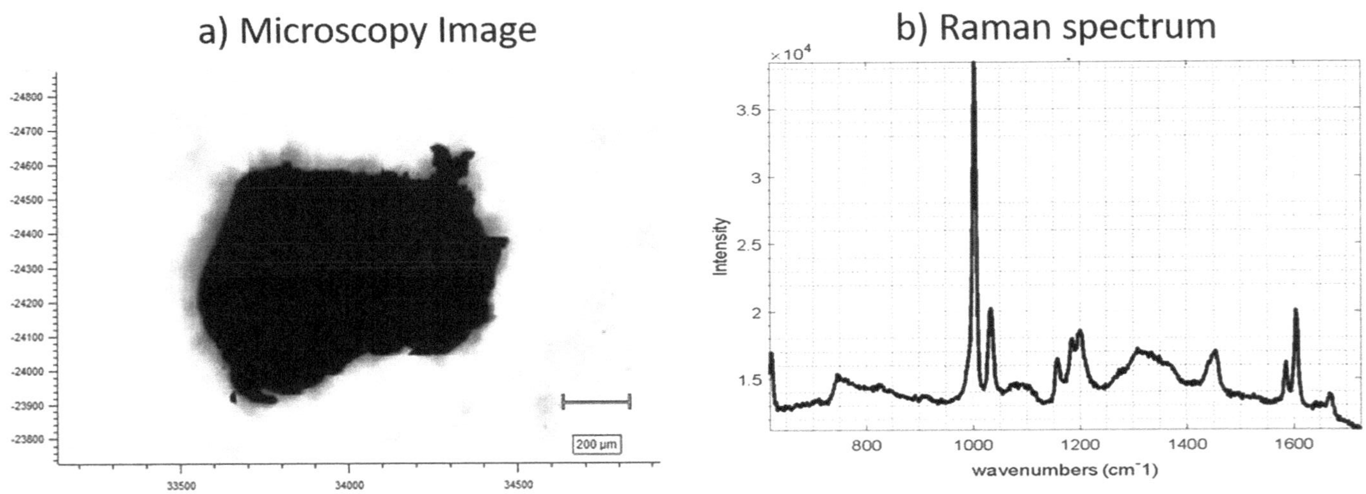

Fig. 5.3 (A) Microscopic image of ABS. (B) Raman spectrum of the ABS microparticle.

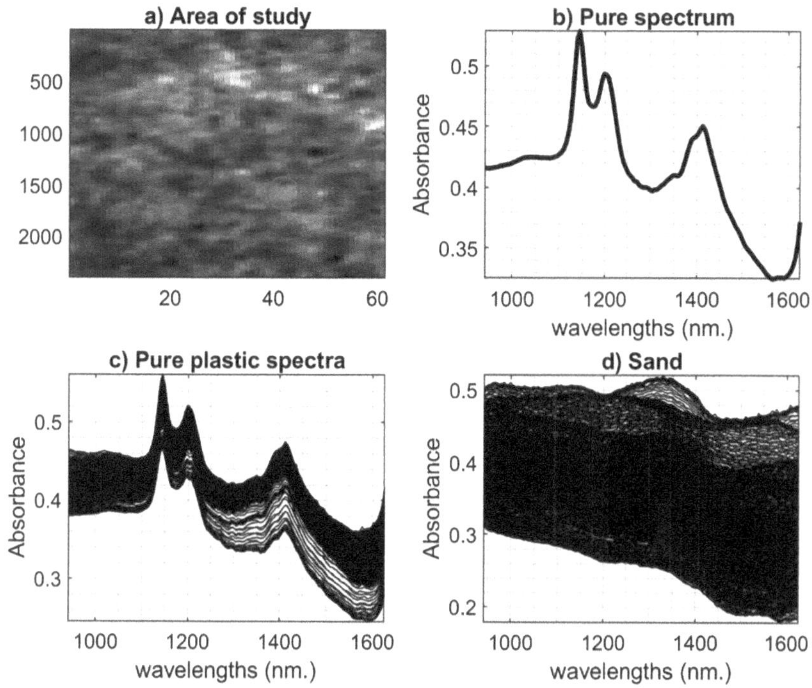

Fig. 5.4 (A) False color image of the petri dish of sand containing ABS microparticle. (B) Average spectrum of the ABS reference pellet. (C) Pure plastic spectra of each pixel containing the ABS reference pellet. (D) The spectra of a region of sand where the spiked ABS microplastic was known not to be located.

Because sand has greater surface roughness than the reference ABS pellets, the extensive scattering of the sand is an unsurprising result. Given that sand is a heterogeneous multicomponent solid matrix with its own NIR fingerprint, there is the potential for interference from the sand in the pixels that contain the ABS microparticle (Amigo & Santos, 2020). Because the sand was burned in an oven at 450°C, there is no trace of organic matter. As such, any interference from the sand would likely arise from silicates and carbonates or many other plausible components that might be found in a complex multicomponent matrix-like sand.

3.3 Exploratory chemometric techniques: PCA

PCA was performed on both SNV and first-derivative spectra with the goal of locating the ABS microparticle (Rinnan, Berg, & Engelsen, 2009). PCA models were calculated with 15 PCs, until the explained variance was lower than 0.001% (Table 5.2). None of the principal components (PCs) of the SNV preprocessed spectra proved useful in locating the ABS microparticle.

The loading of PC1 did not correspond to the spectrum of pure ABS; instead, PC1 has a positive loading in a region where there are no characteristic peaks in the spectrum of ABS (Fig. 5.5). As such, PC1 does not help locate the ABS microparticle. Unlike PC1, PC2 has a positive loading near the characteristic peak of ABS at 1412 nm. However, the majority of the score surface of PC2 shows positive scores, indicating that PC2 is not exclusive and is unable to locate the ABS microparticle. PC3 shows a positive loading near the 1412 nm characteristic peak of ABS and a stronger negative loading in the 1100–1200 nm region, where the 1145-nm and 1198-nm characteristic peaks of ABS are located. The majority of the surface of the PC3 score is negative, with several small groups of pixels with positive scores. Meanwhile, PC4 has a negative loading from 1400–1500 nm near the 1412-nm characteristic peak of ABS. The PC4 scores surface is mostly positive with several small groups of pixels with negative scores. As with the previous principal components, PC3 and PC4 are not exclusive and are unable to locate the ABS microparticle.

Similar to the previously discussed PCA of the SNV preprocessed spectra, principal component analysis of the first derivative preprocessed spectra was not able to locate the ABS microparticle (results not shown).

The outcome of the attempts to locate the ABS microparticle using PCA can reasonably be explained by postulating two major reasons:

(1) The sediment (sand) is a multicomponent matrix where the constituents have a higher reflectivity index than the microplastic. Therefore, the major sources of variance are associated with different components on the surface. These components may be spectrally similar to the spectra of the microplastic. Nevertheless, there is no singular source of

Table 5.2 Percentage of variation in the spectral data explained by each of the first six principal components.

PC	SNV (%)	First derivative (%)
PC1	33.2	22.2
PC2	16.7	16.7
PC3	11.8	9.9
PC4	10.8	8.5
PC5	6.5	7.8
PC6	5.2	6.2

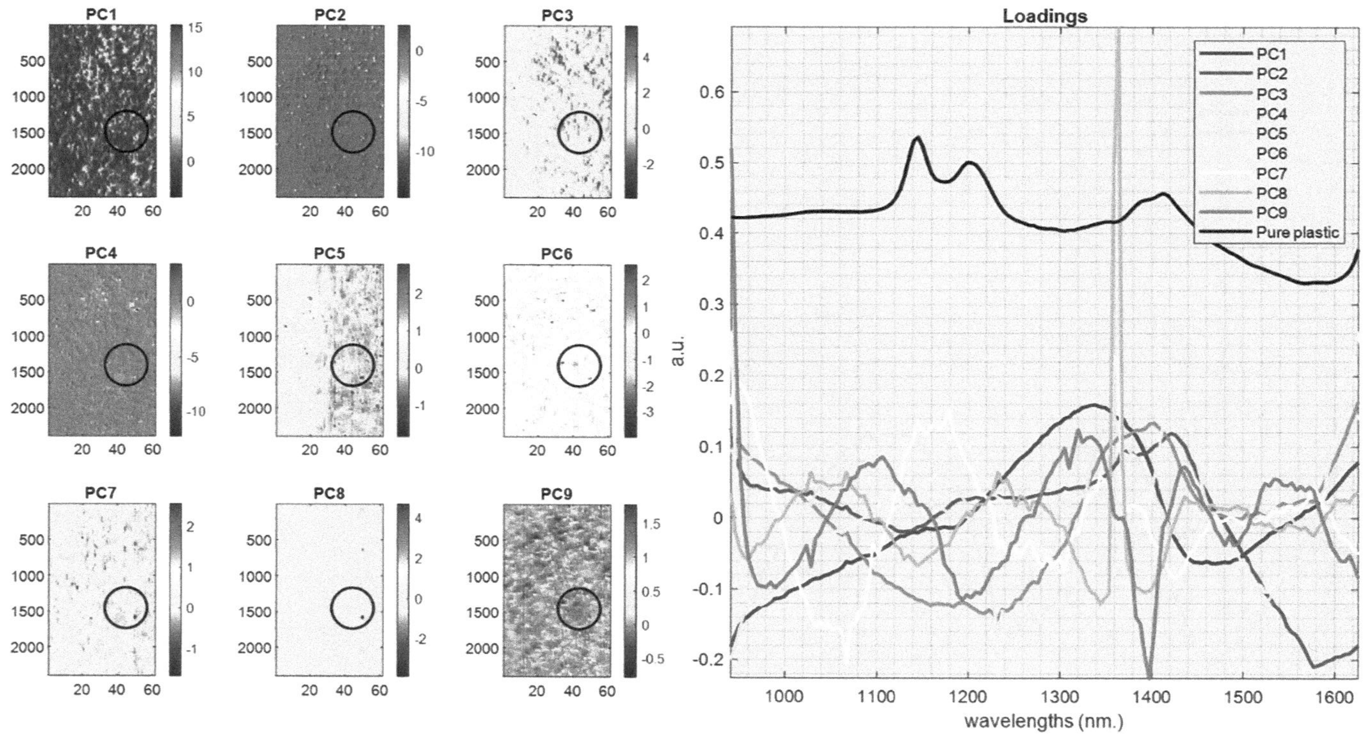

Fig. 5.5 The scores (left) and loadings (right) of the first four principal components of the SNV-preprocessed spectra.

variance that is unequivocally associated with the microplastic. This would be ascribed to a spectral effect.

(2) The spectra associated with the microplastic have been mixed with the spectra of the surrounding elements, meaning that the major microplastic peaks are hidden by the strong influence of the spectra of the sand. This would be ascribed to a spatial effect.

3.4 Classification

Following the failure to locate the ABS microparticle using PCA, classification techniques were then attempted. Classification models sort samples into classes based on their similarity to the measured characteristics of a training set of objects in the hyperspace (Marini, 2010). Recent publications have used SIMCA and PLS-DA as a technique to detect and characterize microplastics with SWIR and FTIR spectroscopy (Da Silva et al., 2020; Serranti et al., 2018; Vidal & Pasquini, 2021).

The results of training both models (with SNV and first-derivative preprocessing) are shown in Table 5.3. Three PCs for the ABS class and four PCs for the sand class were used to build the SIMCA model. PLS-DA models were both built using one single latent variable (LV).

These results clearly indicate the high spectral difference between the sand and the ABS. As is shown further in this document, this is not necessarily a good indication of the performance of the model when pixels from the image are analyzed.

3.5 SIMCA

The SIMCA models were not capable of locating the ABS microparticle in the sand. Neither, the models based on SNV preprocessed data nor that based on the first-derivative preprocessed data succeeded in this regard. Both SIMCA models mischaracterize multiple regions of the sand as ABS under

Table 5.3 Classification results for training and validation sets.

	PLS-DA (SNV)	PLS-DA (first derivative)	SIMCA (SNV)	SIMCA (first derivative)
	TRAIN	VAL	TRAIN	VAL
Sensitivity	100%	100%	100%	100%
Specificity	100%	100%	100%	100%
Error rate	0%	0%	0%	0%

in-class and most probable class (Fig. 5.6). When viewed through the lens of probability, the SIMCA models (both SNV and PLS-DA) indicated a 0% probability that any of the pixels contained the spectrum ABS.

SIMCA is based on PCA of the reference ABS and PCA of the sand class. There is little variation between the spectra that arise from the reference plastic. As such, the spectra of the microplastic need to resemble to a certain extent to the spectra of the reference ABS in order to be classified as such. Sand, meanwhile, is a heterogeneous matrix with a great deal of variation and with its own NIR multiple fingerprints. As we observed in PCA, the spectra of sand influence the spectra of the pixels containing the microplastic and lead to increased differences between their spectra and the reference ABS spectra. The samples, or pixels of the image, in this case, are projected into the PCA spaces of each sand and reference ABS. Because of the influence of sand on the spectra of the pixels containing the microplastic, they are not similar enough to the ABS class to be characterized as in-class.

3.6 PLS-DA

On the contrary to what could be expected, partial least-squares discriminant analysis (PLS-DA) also proved ineffective at locating the ABS microparticle. Eleven clusters of pixels were found to be in-class for ABS by the PLS-DA model based on SNV preprocessed spectra (Fig. 5.7). The map of probability indicates a high probability that each cluster corresponds to the ABS class. As the sample is known to contain one single particle of ABS, this indicates that at least 10 of the clusters represent false positives.

The PLS-DA model based on the first-derivative preprocessed spectra did not successfully locate the ABS microparticle. Instead of finding false positives, this model did not characterize any pixels as ABS.

4. Spectral insights on the problem

The unsuccessful results obtained lead to a more accurate observation of the spectral signatures of the ABS microparticle and sand. Taking advantage of the fact that the approximate location of the microparticle was known, and the spectral peaks of ABS, the hyperspectral image was cropped around the microparticle and the area of the peak between 1100 and 1200 nm was calculated. The results are displayed in Fig. 5.8, left. In this manner, the microparticle was easily identified. The number of pixels highlighted in the X direction of the image is low compared to the Y direction. Therefore, in order to operate better, the hyperspectral

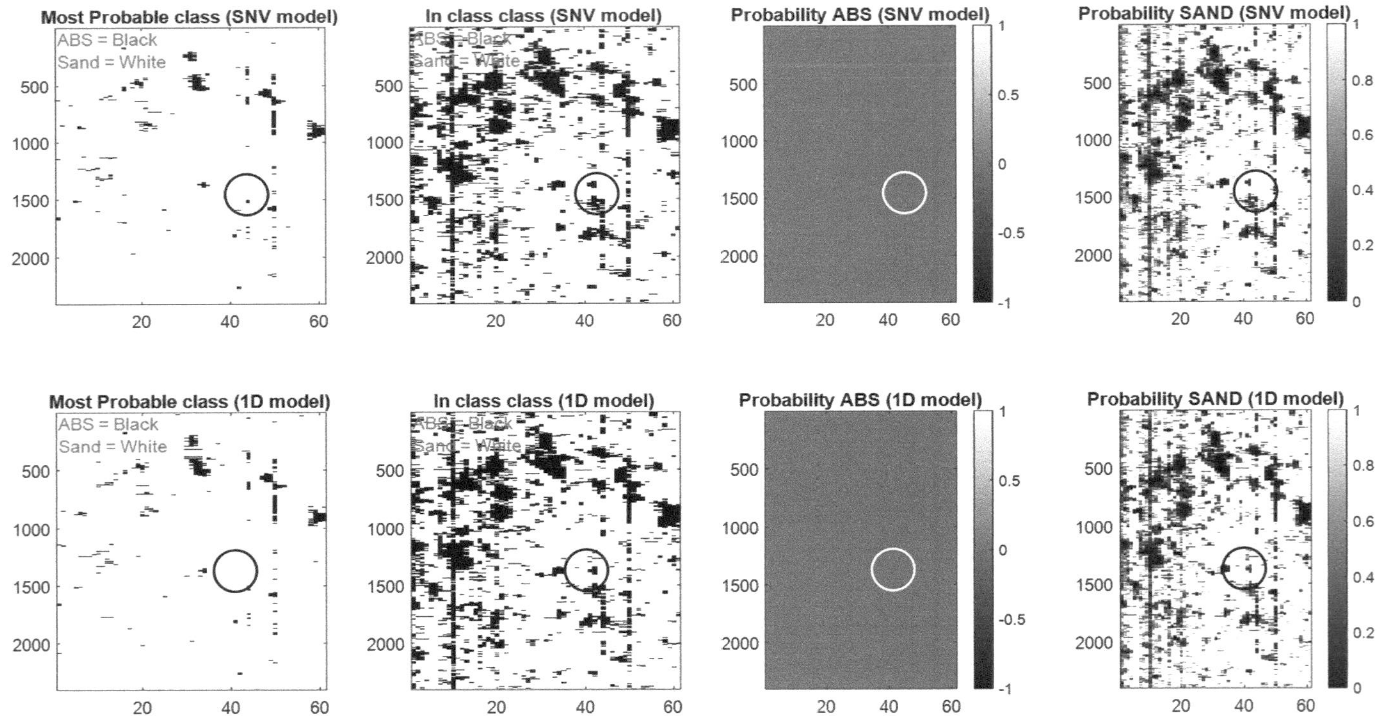

Fig. 5.6 SIMCA model based on SNV preprocessed spectra in the top of the figure, and based on the first-derivative preprocessing in the bottom of the figure.

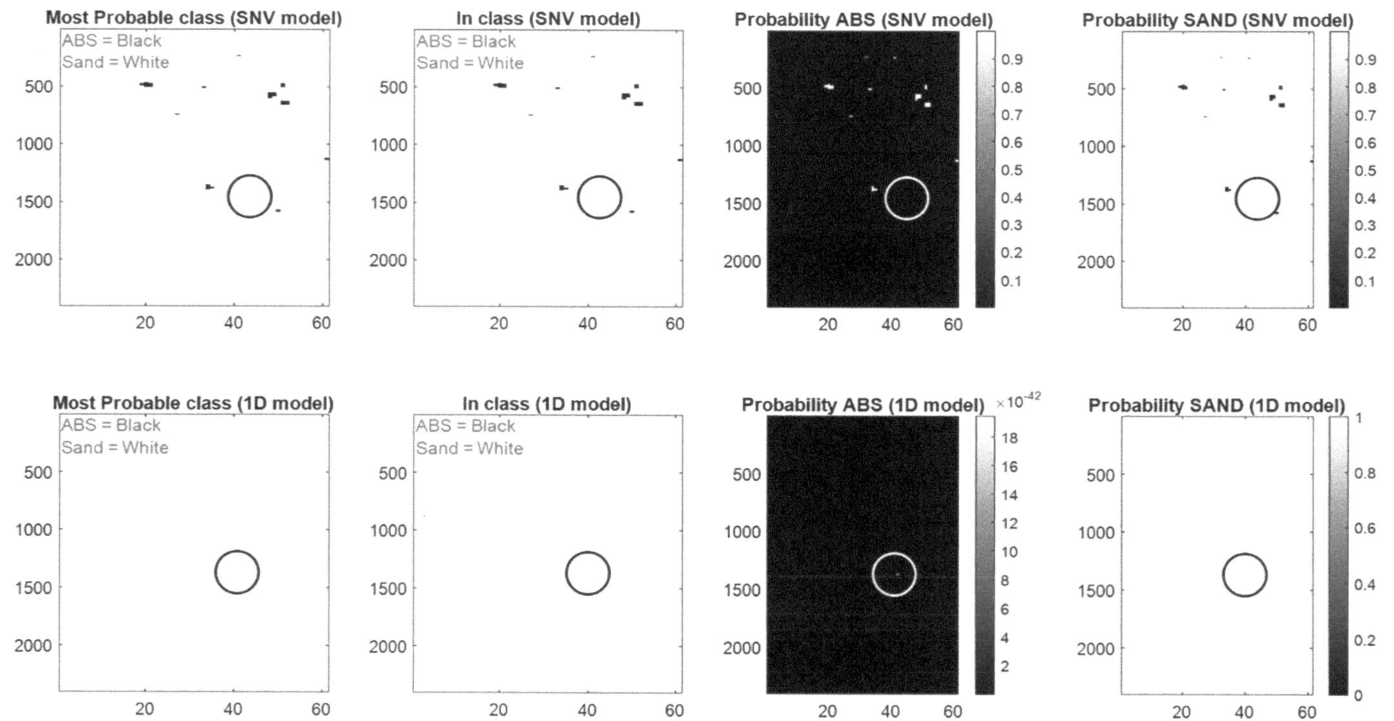

Fig. 5.7 PLS-DA model based on SNV preprocessed spectra in the top of the figure, and based on the first-derivative preprocessing in the bottom of the figure.

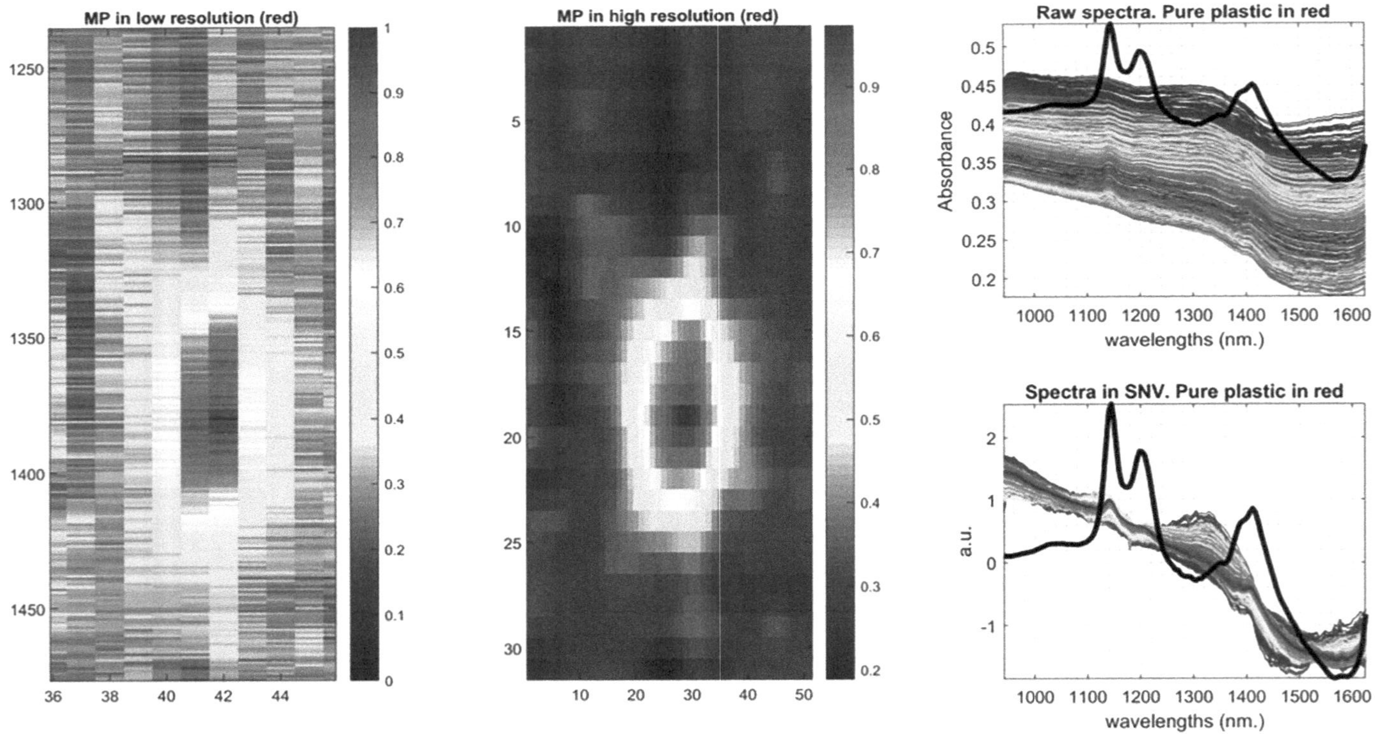

Fig. 5.8 Left, normalized area of the peak between 1100 and 1200 nm. Middle, a resized hyperspectral image of the same area. Right, spectral signatures (raw and SNV) of the spectra colored according to the color degree seen in the figure in the middle.

image was rescaled in both directions (using bicubic interpolation). By doing so, the spectral resolution is not affected, since the neighbor pixels are mere interpolations of the real pixels. Nevertheless, a better spatial resolution is obtained (Fig. 5.8).

Observing the raw spectra of the selected area (Fig. 5.8), it can be observed how the scattering of the pixels containing sand is much larger than the scattering of the pixels containing any trace of ABS spectrum. When the scattering is corrected (with SNV), it looks evident that the signal of the ABS in the pixels belonging to the microparticle is several degrees smaller than the signal of the pure plastic. The spectra of sand influence the spectra of the pixels containing the microplastic and leading to increased differences between their spectra and those of reference ABS.

These observations, together with the previous results, point to answers to the questions posted at the very beginning of this chapter. Not only does the chemometric strategy play a fundamental role. Instead, it is of utmost importance to be aware of the effects of the chemometric strategy, the nature of the microparticle, the nature of the background, and the suitability of the instrument:

- The instrument used in this study might not be the most adequate for this task. Better said, the limit of detection (spatial and spectral) of the instrument was not good enough.
- It might also be that the chemometric strategies were not the most adequate ones. This opens up endless space for improvement. Nevertheless, the chemometric strategy is strongly linked to the database quality that is used. In this case, the strategy used for building the database is the most common strategy. This suggests that any spectrum in the sample should resemble to a certain degree to the spectra in the library.
- It has been demonstrated that the multicomponent character of the sand (surroundings of the microparticle) also plays a fundamental role, especially if the NIR beam penetrates the microparticle and if the reflection intensities of the sand are much larger than the reflection intensity of the microparticle.
- The microparticle is composed by the same material as the pellets used for building the database. Moreover, microparticles were measured at the same time and under the same conditions. Nevertheless, the spectrum of the microparticle is much smaller than the spectrum of the pure plastic. This demonstrates that there is a strong interaction between the spectra of the microparticle and the scattering effects of the spectra of the sand.

5. Future work

NIR-HSI is gaining traction as a feasible technique for the detection and classification of microplastics directly in sand, soil, and marine sediments, which is an exciting prospect given the labor-intensive nature of separating microplastics from heterogeneous solid matrices. Given that this is a relatively new application of NIR-HSI, it is important to understand the limits and uncertainty associated with it. The results obtained clearly indicate that:

(1) The NIR radiation penetrates the microparticle, suggesting a "transflectance" effect between the sand and the plastic. This not only amplifies the signal of the microplastic, but also amplifies the signal of the sand, which appears to have high reflectivity power. The small size of the microparticle, together with its low thickness, suggests that even the spectra located in the middle of the microparticle suffer from the high influence of the spectra of the sand.

(2) The spectra of the edges of the microplastic are certainly highly affected by the scattering of the sand. This suggests uncertainty in the characterization of those pixels as microplastics, creating an uncertainty boundary that must be further studied.

As said before, this work is an introduction to a more complex topic where different theories and algorithms will be tested to calculate the spatial and spectral limits of detection.

From a more general perspective, these results call into question the strategy of constructing libraries containing only the spectra of pure plastic and sand (or different varieties of sand). This chapter demonstrates that constructing classification models by using standardized signals might create too narrow of boundaries in the classification models, impeding their ability to find microplastics present in environmental media.

This last theory presented here about the limits of detection in the spatial and spectral domains requires further study. The proposal is not finished, since there are many aspects that should be considered. Concepts like net analyte signal and linearity of NIR radiation will be evaluated to fully understand the capabilities of NIR-HSI and to assess the extent to which NIR-HSI, together with chemometrics, is able to give an accurate and reliable answer.

Acknowledgments

This work has been developed with the support of the Spanish Ministry of Science and Innovation through projects PID2020-118685RB-I00 and ELKARTEK21/50 Investigación estratégica de las propiedades, aplicaciones y efectos de los nanomaterials poliméricos.

References

Amigo, J. M. (2020). Hyperspectral and multispectral imaging: Setting the scene. In *Vol. 32. Data handling in science and technology* (pp. 3–16). Elsevier Ltd. https://doi.org/10.1016/B978-0-444-63977-6.00001-8.

Amigo, J. M., Babamoradi, H., & Elcoroaristizabal, S. (2015). Hyperspectral image analysis. A tutorial. *Analytica Chimica Acta, 896*, 34–51. https://doi.org/10.1016/j.aca.2015.09.030.

Amigo, J. M., & Santos, C. (2020). Preprocessing of hyperspectral and multispectral images. In *Vol. 32. Data handling in science and technology* (pp. 37–53). Elsevier Ltd. https://doi.org/10.1016/B978-0-444-63977-6.00003-1.

Arthur, C., & Baker, H. (2008). *Proceedings of the international research workshop on microplastic marine debris | OR&R's marine debris program.*

Barnes, D. K. A., Galgani, F., Thompson, R. C., & Barlaz, M. (2009). Accumulation and fragmentation of plastic debris in global environments. *Philosophical Transactions of the Royal Society, B: Biological Sciences, 364*(1526), 1985–1998. https://doi.org/10.1098/rstb.2008.0205.

Bro, R., & Smilde, A. K. (2014). Principal component analysis. *Analytical Methods, 6*(9), 2812–2831. https://doi.org/10.1039/c3ay41907j.

Cressey, D. (2016). Bottles, bags, ropes and toothbrushes: The struggle to track ocean plastics. *Nature, 536*(7616), 263–265. https://doi.org/10.1038/536263a.

Da Silva, V. H., Murphy, F., Amigo, J. M., Stedmon, C., & Strand, J. (2020). Classification and quantification of microplastics (<100 μm) using a focal plane Array-Fourier transform infrared imaging system and machine learning. *Analytical Chemistry, 92*(20), 13724–13733. https://doi.org/10.1021/acs.analchem.0c01324.

Dong, M., Zhang, Q., Xing, X., Chen, W., She, Z., & Luo, Z. (2020). *A Raman database and the surface changes of microplastics weathered under natural environments.* https://doi.org/10.13140/RG.2.2.10929.74086.

Eriksen, M., Lebreton, L. C. M., Carson, H. S., Thiel, M., Moore, C. J., Borerro, J. C., et al. (2014). Plastic pollution in the World's oceans: More than 5 trillion plastic pieces weighing over 250,000 tons afloat at sea. *PLoS One, 9*(12), e111913. https://doi.org/10.1371/journal.pone.0111913.

Frias, J. P. G. L., & Nash, R. (2019). Microplastics: Finding a consensus on the definition. *Marine Pollution Bulletin, 138*, 145–147. https://doi.org/10.1016/j.marpolbul. 2018.11.022.

Frias, J. P. G. L., Pagter, E., Nash, R., O'Connor, Carretero, O., Filgueiras, A., et al. (2018). *Standardised protocol for monitoring microplastics in sediments.* https://doi.org/10.13140/RG.2.2.36256.89601/1.

Geyer, R., Jambeck, J. R., & Law, K. L. (2017). Production, use, and fate of all plastics ever made. *Science Advances, 3*(7). https://doi.org/10.1126/sciadv.1700782.

Guidance on Monitoring of Marine Litter in European Seas. (2013). *Joint research center scientific and policy reports.*

Halsband, C., & Herzke, D. (2019). Plastic litter in the European Arctic: What do we know? *Emerging Contaminants, 5*, 308–318. https://doi.org/10.1016/j.emcon.2019.11.001.

Hidalgo-Ruz, V., Gutow, L., Thompson, R. C., & Thiel, M. (2012). Microplastics in the marine environment: A review of the methods used for identification and quantification. *Environmental Science and Technology, 46*(6), 3060–3075. https://doi.org/10.1021/es2031505.

Horton, A. A., Walton, A., Spurgeon, D. J., Lahive, E., & Svendsen, C. (2017). Microplastics in freshwater and terrestrial environments: Evaluating the current understanding to identify the knowledge gaps and future research priorities. *Science of the Total Environment, 586*, 127–141. https://doi.org/10.1016/j.scitotenv.2017.01.190.

Käppler, A., Fischer, D., Oberbeckmann, S., Schernewski, G., Labrenz, M., Eichhorn, K. J., et al. (2016). Analysis of environmental microplastics by vibrational microspectroscopy:

FTIR, Raman or both? *Analytical and Bioanalytical Chemistry*, *408*(29), 8377–8391. https://doi.org/10.1007/s00216-016-9956-3.

Kershaw, P. (2015). *Sources, fate and effects of microplastics in the marine environment: A global assessment*. International Maritime Organization.

Kershaw, & Rochman, C. M. (2015). *Sources, fate and effects of microplastics in the marine environment: Part 2 of a global assessment*. Joint Group of Experts on the Scientific Aspects of Marine Environmental Protection.

Lefebvre, C., Rojas, I. J., Lasserre, J., Villette, S., Lecomte, S., Cachot, J., et al. (2021). Stranded in the high tide line: Spatial and temporal variability of beached microplastics in a semi-enclosed embayment (Arcachon, France). *Science of the Total Environment*, *797*, 149144. https://doi.org/10.1016/j.scitotenv.2021.149144.

Marini, F. (2010). Classification methods in chemometrics. *Current Analytical Chemistry*, *6*(1), 72–79. https://doi.org/10.2174/157341110790069592.

Massart, D. L. (1997). Handbook of chemometrics and qualimetrics (1st ed.). Elsevier.

Mobaraki, N., & Amigo, J. M. (2018). HYPER-tools. A graphical user-friendly interface for hyperspectral image analysis. *Chemometrics and Intelligent Laboratory Systems*, *172*, 174–187. https://doi.org/10.1016/j.chemolab.2017.11.003.

Plastics Europe. (2020). *Plastics—The facts 2020: An analysis of European plastics production, demand and waste data*. PlasticsEurope Association of Plastic Manufacturers.

Ragusa, A., Svelato, A., Santacroce, C., Catalano, P., Notarstefano, V., Carnevali, O., et al. (2021). Plasticenta: First evidence of microplastics in human placenta. *Environment International*, *146*. https://doi.org/10.1016/j.envint.2020.106274.

Rinnan, A., Berg, F. V. D., & Engelsen, S. B. (2009). Review of the most common preprocessing techniques for near-infrared spectra. *TrAC, Trends in Analytical Chemistry*, *28*(10), 1201–1222. https://doi.org/10.1016/j.trac.2009.07.007.

Ruggero, F., Gori, R., & Lubello, C. (2020). Methodologies for microplastics recovery and identification in heterogeneous solid matrices: A review. *Journal of Polymers and the Environment*, *28*(3), 739–748. https://doi.org/10.1007/s10924-019-01644-3.

Serranti, S., Palmieri, R., Bonifazi, G., & Cózar, A. (2018). Characterization of microplastic litter from oceans by an innovative approach based on hyperspectral imaging. *Waste Management*, *76*, 117–125. https://doi.org/10.1016/j.wasman.2018.03.003.

Vidal, C., & Pasquini, C. (2021). A comprehensive and fast microplastics identification based on near-infrared hyperspectral imaging (HSI-NIR) and chemometrics. *Environmental Pollution*, *285*, 117251. https://doi.org/10.1016/j.envpol.2021.117251.

Xu, S., Ma, J., Ji, R., Pan, K., & Miao, A.-J. (2020). Microplastics in aquatic environments: Occurrence, accumulation, and biological effects. *Science of the Total Environment*, *703*, 134699. https://doi.org/10.1016/j.scitotenv.2019.134699.

Yang, L., Zhang, Y., Kang, S., Wang, Z., & Wu, C. (2021). Microplastics in soil: A review on methods, occurrence, sources, and potential risk. *Science of the Total Environment*, *780*, 146546. https://doi.org/10.1016/j.scitotenv.2021.146546.

CHAPTER 6

Autoencoders in generative modeling, feature extraction, regression, and classification

Mohsen Namazi[a], Mohammad Hossein Karimi-Jafari[a], Farzad Qassemi[c], and Jahan B. Ghasemi[b]
[a]Department of Bioinformatics, Institute of Biochemistry and Biophysics, University of Tehran, Tehran, Iran
[b]Faculty of Chemistry, University of Tehran, Tehran, Iran
[c]AIQTech Inc, Toronto, ON, Canada

1. What is an autoencoder?

An *autoencoder* is a two-part machine. The first part, the *encoder*, is a function f (deterministic or stochastic) that maps input $\boldsymbol{x}$ to output $\boldsymbol{z}$. So, $\boldsymbol{z}=f(\boldsymbol{x})$. In this chapter, we call $\boldsymbol{z}$ as *latent representation* or *code*. The second part, the *decoder*, is a function g (also deterministic or stochastic) that maps input $\boldsymbol{z}$ to output $\boldsymbol{r}$. Therefore, $\boldsymbol{r}=g(\boldsymbol{z})$. We want $g(f(\boldsymbol{x}))$ to be as similar as possible to input $\boldsymbol{x}$. In other words, function g attempts to reconstruct $\boldsymbol{x}$, from input $\boldsymbol{z}$. Applying some restrictions on f and g, we prevent complete copying and so the machine is forced to find the most relevant features (Goodfellow, Bengio, & Courville, 2016). Theoretically speaking, f and g can be from an arbitrary functional family, but *neural networks* (e.g., feed-forward neural networks) are of most interest. Different types of autoencoders have been proposed, which we will review in the following subsections.

1.1 Undercomplete autoencoders

In this type of autoencoders, the latent space dimensions are less than the input space dimensions, and they are generally used to reduce dimensionality, feature extraction, and compression. Latent space representation has the essential input characteristics as much as possible. Fig. 6.1 shows the architecture of this type of network.

1.2 Overcomplete autoencoders

Generally, when using dimensionality reduction techniques, there is an implicit assumption: data can be embedded in a smaller space. However,

Machine Learning and Pattern Recognition Methods in Chemistry from Multivariate and Data Driven Modeling
https://doi.org/10.1016/B978-0-323-90408-7.00007-1

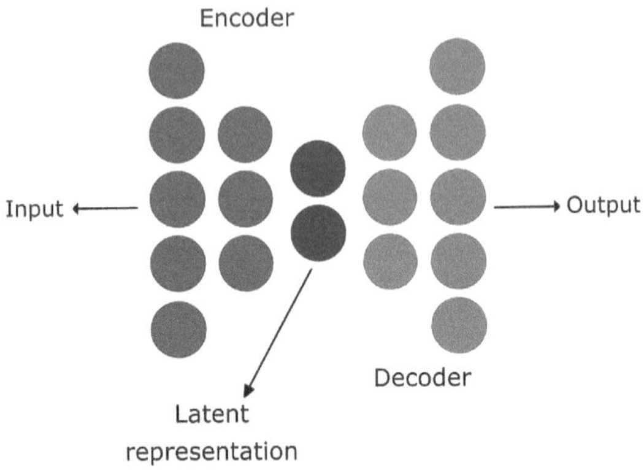

Fig. 6.1 Undercomplete autoencoder architecture. In undercomplete autoencoders, latent space dimensions are less than the input space dimensions.

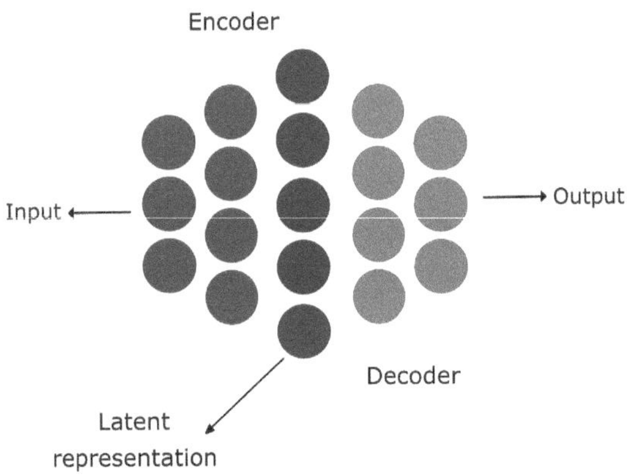

Fig. 6.2 Overcomplete autoencoder architecture. In overcomplete autoencoders, latent space dimensions are larger than the input space dimensions.

it has been shown that a valuable representation of the data can be found with the opposite assumption (Ng et al., 2011). Of course, regularization methods usually have to be used for this purpose. We will have an overcomplete autoencoder if the latent space dimensions are larger than the input space dimensions. Fig. 6.2 shows the architecture of this type of network.

1.3 Stacked autoencoder

Training neural networks in a greedy layer-wise manner were done first by Hinton and Salakhutdinov (2006). The same idea has been used in *stacked autoencoder*. In general, this algorithm can be described as follows:

1. Each layer of the network is trained to extract information as much as possible.

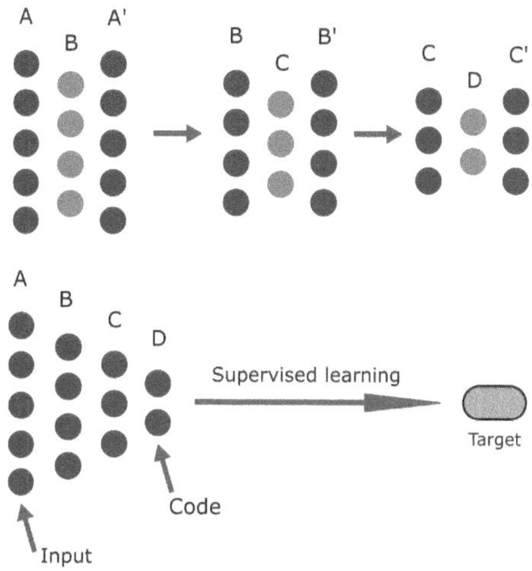

Fig. 6.3 Stacked autoencoder training process. Stacked autoencoder is trained in a greedy layer-wise manner.

2. Resorting layers (i.e., A', B', C' in Fig. 6.3) are discarded in the training process.
3. The weights of the hidden layers (i.e., A, B, C in Fig. 6.3) are considered fixed and practically considered the input of the next layer.
4. All the hidden layers are combined, and the whole network is created.
5. A fine-tuning operation is used to improve the network performance.

1.4 Sparse autoencoder

The main idea of sparsity was first introduced in computational neuroscience (Olshausen & Field, 1997) and later used in machine learning. In overcomplete architecture, the code consists of a large number of nodes. However, the activation value of these nodes in most instances is zero. In other words, the average activation value for each node is close to zero. Finding a representation for the input instances that extracts only the essential features is the main logic of this approach. To train a sparse autoencoder, we use an objective function that includes the regularization penalty:

$$L(\boldsymbol{x}, g(f(\boldsymbol{x}))) + \Omega(\boldsymbol{z}) \tag{6.1}$$

where $g(\boldsymbol{z})$ is the decoder output, $\boldsymbol{z} = f(\boldsymbol{x})$ is the encoder output or code, and $\Omega(\boldsymbol{z})$ is a sparsity penalty. The effect of using the sparsity component on the

MNIST dataset has been evaluated (Makhzani & Frey, 2014). As the intensity of regularization increases, the model attempts to find features that provide a complete representation of digit. If the intensity of regularization decreases, the model focuses on local features.

1.5 Denoising autoencoder

The idea of adding noise to the dataset was probably first introduced in Lecun (1987) and Soulie, Gallinari, Le Cun, and Thiria (1987). It was then also used in convolutional and recurrent neural networks. Finally, Vincent et al. (2010) proposed a way to use this approach for training autoencoders.

The denoising autoencoder is trained to detect and eliminate noises. So noise is added to the instances, and the network learns how to reconstruct the noise-free input samples, and in this process, the most robust and crucial features are extracted. The input in this case is a corrupted version $\widetilde{\boldsymbol{x}} \in \mathbb{R}^n$ of $\boldsymbol{x}$.

The loss function takes the following form:

$$L(\boldsymbol{x}, g(f(\widetilde{\boldsymbol{x}}))) \tag{6.2}$$

where $g(f(\widetilde{\boldsymbol{x}}))$ is the decoder output, and $f(\widetilde{\boldsymbol{x}})$ is the encoder output encoded from the corrupted input $\widetilde{\boldsymbol{x}}$ (see Fig. 6.4).

1.6 Convolutional autoencoder

The *convolution operator* is the central part of this network. Convolutional networks were first proposed by Fukushima and Miyake (1982) but modern

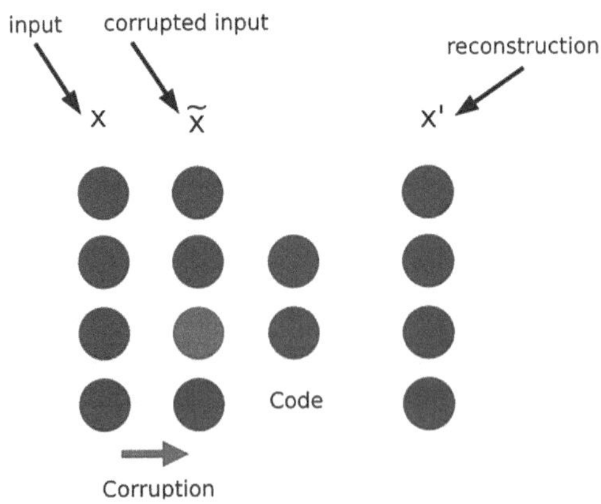

Fig. 6.4 Denoising autoencoder. Denoising autoencoder learns how to reconstruct original input from the corrupted version.

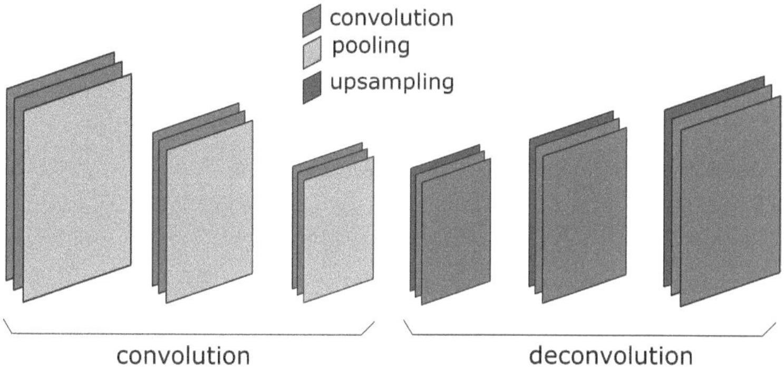

Fig. 6.5 Convolutional autoencoder. Typical use of convolutional autoencoder is feature extraction.

formulation can be found in LeCun et al. (1989) and LeCun, Bottou, Bengio, and Haffner (1998). In this type of architecture, the implicit assumption is that the data has a grid structure or has a proper matrix or tensor representation such that the proximity of elements increases their correlation.

Two main parts of convolutional networks are convolutional and subsampling layers. Usually, these layers are followed by a fully connected layer. Feature maps are extracted from raw input representation. Parameters in this type of operation are shared across the entire input tensor (or matrix). There is another mechanism, called pooling, that reduces the dimensionality of the feature map. Pooling maps values in small regions to a single value (e.g., maximum or average value). This is considered as one typical CNN layer. It is possible to use more layers to increase the complexity of the model. Multiple CNN layers help high-level features detection. Usually, there is a fully connected layer at the end of the network (Kazak, 2019).

In classification or regression, the fully connected layer helps to train a predictive model that uses high-level features extracted through CNN layers. However, in unsupervised learning, up-sampling layers are applied to reconstruct the input tensor (Kazak, 2019).

Typical use of convolutional autoencoder is feature extraction. However, it is common to use for pretraining convolutional neural networks (Kohlbrenner, Hofmann, Ahmmed, & Kashef, 2017) (see Fig. 6.5).

1.7 Sequence-to-sequence autoencoder

One crucial data type is sequence data. In molecular biology, DNA or protein sequences are of most interest. Also, developing language models has lots of applications. Recurrent neural networks (RNNs) have been developed for sequence modeling. Sequence-to-sequence autoencoder is based on RNN. There are lots of RNN variants. The most famous architectures are long short-term memory (LSTM) (Hochreiter & Schmidhuber, 1997), gated recurrent unit (GRU) (Chung, Gulcehre, Cho, & Bengio, 2014), and bidirectional RNN (Schuster & Paliwal, 1997).

When we want to model a time series, generally system's behavior is dependent on past events, so the model should have some memory. RNN architectures provide a mechanism for memorizing valuable information across sequences. In these models, there is a recurrent formula that updates the system state at each time step.

$$h_t = f_W\,(h_{t-1}, x_t) \tag{6.3}$$

$$y_t = W_{hy}\, h_t \tag{6.4}$$

where h_t is a new state, f_W is a function with parameters W, h_{t-1} is an old state, x_t is an input vector at some time step, and y_t is a sequence of outputs.

Historically speaking, recurrent neural nets had a significant obstacle in the training process. When the network's output is fed back to the input, vanishing gradient problem occurs. Long short-term memory (LSTM) (Hochreiter & Schmidhuber, 1997) network was a solution for this problem. This architecture uses a gating mechanism for managing this difficulty.

In natural language processing, RNNs are widely used (Bengio, Ducharme, Vincent, & Janvin, 2003; Socher, Huang, Pennin, Manning, & Ng, 2011). Also, in the machine translation problem, a sequence-to-sequence (Seq2Seq) model was used (Kohlbrenner et al., 2017).

In this model, the main idea is training an encoder that learns a continuous representation of words and sentences. This representation conserves the semantic and syntactic structure of source language as much as feasible. The decoder maps this representation to equivalent words and sentences in the target language (Cho et al., 2014; Sutskever, Vinyals, & Le, 2014) (see Fig. 6.6).

1.8 Variational autoencoders

Variational autoencoders are the marriage of *directed probabilistic graphical models* and neural networks. An effective method for understanding the general mechanisms of a system is using *probabilistic modeling*. These models formalize the notion of knowledge in a mathematically rigorous manner. Generally speaking, in probabilistic viewpoint, we allow a controlled level of ambiguity and randomness in the model variables. Using the concept of conditional

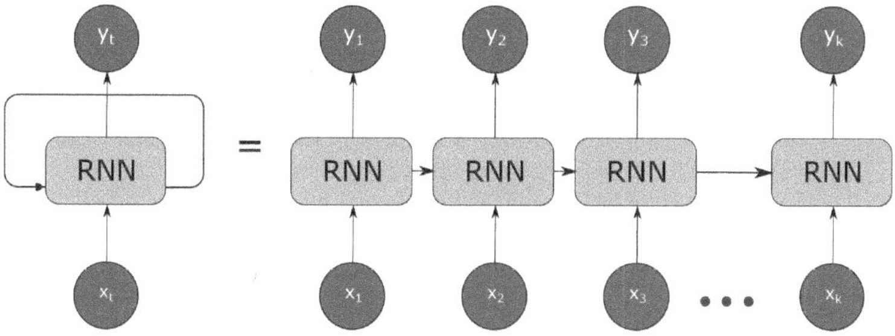

Fig. 6.6 Recurrent neural network unfolding. The main idea of recurrent neural networks is sharing parameters.

probability distribution, the level and nature of this uncertainty can be explicitly described. In a *Bayesian probabilistic modeling* framework, we set a *prior probability distribution* over the model's unknown parameters (latent variables). After seeing data, we update this probability distribution to *posterior probability distribution*. One effective method for computing such a posterior is using *variational inference*. Following descriptions are mainly from Kingma (2017).

1.9 Probabilistic modeling and neural networks

If the probability distribution over model variables be available, then understanding system behaviors are mainly accessible. We represent model variables with the letter $\boldsymbol{x}$. So, the main goal in probabilistic modeling is finding $p(\boldsymbol{x})$, the joint probability distribution of model variables. Furthermore, we introduce a process that is applicable for approximating $p^*(\boldsymbol{x})$, with $p_\theta(\boldsymbol{x})$ such that:

$$p_\theta(\boldsymbol{x}) \approx p^*(\boldsymbol{x}) \tag{6.5}$$

Also, it is common to approximate a *conditional probability distribution*. For example, in supervised learning (e.g., classification and regression), $\boldsymbol{x}$ is the model's input, and we want to predict y, the model's output. From a probabilistic point of view, we want to approximate $p(y|\boldsymbol{x})$.

After taking this probabilistic framework, the next step is introducing a functional family for parameterizing the probability distributions. In variational autoencoders, the choice is neural networks. Generally, a neural network is a vector function, NeuralNet(.). For instance, in image classification, if we represent the image with the letter $\boldsymbol{x}$ and the class label with the letter y, then

$$\boldsymbol{p} = \text{NeuralNet}(\boldsymbol{x}) \tag{6.6}$$

$$p_\theta\ (y|\boldsymbol{x}) = \text{Categorial}(y; \boldsymbol{p}) \tag{6.7}$$

where usually, in the NeuralNet, the last operation is softmax() function such that $\sum_i p_i = 1$. If we define a *directed acyclic graph* that expresses conditional dependence structure between random variables, the probability distribution over $\boldsymbol{x} = (x_1, x_2, \ldots, x_M)$ can be factorized as

$$p_\theta(x_{1,\ldots,}x_M) = \prod_{i=1}^{M} p_\theta(x_i|pa(x_i)) \tag{6.8}$$

where $Pa(x_i)$ is the set of parent variables of node i in the directed graph. For root nodes, the set of parents is empty set and the distribution is unconditional. This is a *directed probabilistic graphical model* or *Bayesian network*. A flexible way to parameterize such distributions is with neural networks. The input for this neural network is the vector of parent variables and the output is parameters $\boldsymbol{\mu}$ of probability distribution over child variable $\boldsymbol{x}$:

$$\boldsymbol{\mu} = \text{NeuralNet}(Pa(\boldsymbol{x})) \tag{6.9}$$

$$p_\theta\ (\boldsymbol{x}|Pa(\boldsymbol{x})) = p_\theta\ (\boldsymbol{x}|\boldsymbol{\mu}) \tag{6.10}$$

1.10 Learning and inference problem

In probabilistic modeling, it is common to present some *latent (hidden) variables*. Latent variables, as their name suggests, are not observed in the dataset. We represent these variables with letter $\boldsymbol{z}$. So, the directed probabilistic graphical model should factorize $p_\theta\ (\boldsymbol{x}, \boldsymbol{z})$. For computing probability distribution over $\boldsymbol{x}$, using marginalization, we have:

$$p_\theta(\boldsymbol{x}) = \int p(\boldsymbol{x},\ \boldsymbol{z})d\boldsymbol{z} \tag{6.11}$$

For example, if $\boldsymbol{z}$ is discrete multinomial and also $p_\theta\ (\boldsymbol{x} \mid \boldsymbol{z})$ is normal distribution, then $p_\theta(\boldsymbol{x})$ is a mixture of Gaussians distribution. Usually, the latent variable model $p_\theta(\boldsymbol{x}, \boldsymbol{z})$, whose distributions are parameterized by neural networks, is called *deep latent variable model* (DLVM). We also use this term. This model can be conditioned (e.g., $p_\theta(\boldsymbol{x}, \boldsymbol{z} \mid y)$). One interesting feature of DLVMs is that it is possible to model complicated $p_\theta(\boldsymbol{x})$ using simple factors (conditional distributions). As an example, suppose latent variables with Gaussian and $\boldsymbol{x} \mid \boldsymbol{z}$ with Bernoulli distributions:

$$p_\theta(\boldsymbol{z}) = N(\boldsymbol{z}; \boldsymbol{0}, \boldsymbol{I}) \tag{6.12}$$

$$p = \text{DecoderNeuralNet}_\theta(z) \tag{6.13}$$

$$\log p_\theta(x|z) \tag{6.14}$$

where $\forall p_j \in \boldsymbol{p} : 0 \leq p_j \leq 1$ (e.g., using Sigmoid(.) for the last layer of $\text{DecoderNeuralNet}_\theta$ (.). *Maximum-likelihood estimation* can be applied for determining proper model parameters. Though, commonly computing p_θ $(\boldsymbol{x})$ is intractable. The reason is that the integral in equation above has no analytic or efficient approximate solution necessarily. So, we cannot differentiate it w.r.t. its parameters and run a gradient-based optimization algorithm. In the equation $p_\theta\ (\boldsymbol{z}|\boldsymbol{x}) = \frac{p_\theta(\boldsymbol{x},\ \boldsymbol{z})}{p_\theta(\boldsymbol{x})}$, computing joint probability distribution, $p_\theta(\boldsymbol{x}, \boldsymbol{z})$, is efficient. Therefore, both posterior, $p_\theta(\boldsymbol{z}\,|\,\boldsymbol{x})$, and marginal probability, $p_\theta(\boldsymbol{x})$, are intractable. There are some approximate inference techniques for computing $p_\theta(\boldsymbol{z}\,|\,\boldsymbol{x})$ and $p_\theta(\boldsymbol{x})$, but they are not scalable enough. For example, they need a per-datapoint optimization loop or the posterior approximation is not precise. So, the main question is:

How to perform a scalable and efficient approximate posterior or maximum-likelihood estimation inference in deep latent variable models?

1.11 Efficient variational inference

The strategy for handling interactability problems in DLVM inference and learning is based on first-order gradients w.r.t parameters and latent variables. Then, backpropagation is efficient for finding solutions to this optimization problem.

Variational autoencoder (VAE) framework combines an encoder neural net for inference with a decoder neural net for generative modeling. Approximate solutions for this model will be found by optimizing an objective function. The objective function is a lower bound for the log-likelihood of the parameters given data. *Doubly stochastic gradient descent* provides the scalability of this procedure. The variational inference model, q_ϕ $(\boldsymbol{z}\,|\,\boldsymbol{x})$, is an approximation for p_θ $(\boldsymbol{z}\,|\,\boldsymbol{x})$ (see Fig. 6.7).

where ϕ is the vector of variational parameters. Also, a factorization for $q_\phi(\boldsymbol{z}\,|\,\boldsymbol{x})$ is available (similar to $p_\theta(\boldsymbol{z}\,|\,\boldsymbol{x})$ factorization):

$$q_\phi(\boldsymbol{z}|\boldsymbol{x}) = q_\phi(z_1, z_{2,\ldots,}z_M|\boldsymbol{x}) = \prod_{i=1}^{M} q_\phi(z_i|P_a(z_i), \boldsymbol{x}) \tag{6.15}$$

q_ϕ $(\boldsymbol{z}\,|\,\boldsymbol{x})$ is parameterized with a neural network. In other words, parameters ϕ include the weights and biases of this neural net. As an example:

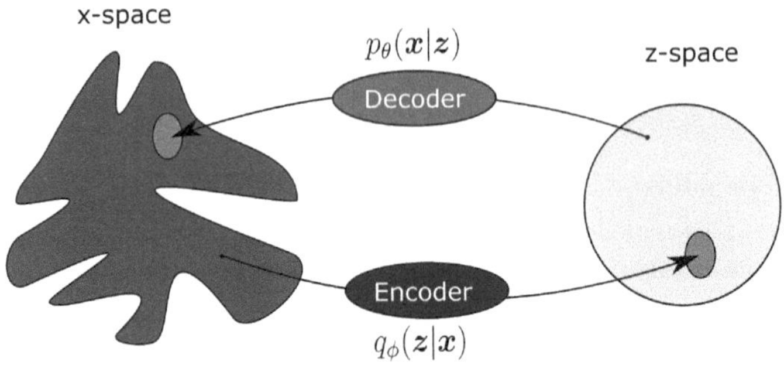

Fig. 6.7 Variational autoencoder. Variational autoencoder is a combination of two stochastic mappings.

$$(\boldsymbol{\mu}, \log s,) = EncoderNeuralNet_{\phi}\,(\boldsymbol{x}) \tag{6.16}$$

$$q_{\phi}\,(\boldsymbol{z}|\boldsymbol{x}) = N(\boldsymbol{z}; \boldsymbol{\mu}, diag(s)) \tag{6.17}$$

The optimization objective of the variational autoencoder is the *evidence lower bound*, abbreviated as *ELBO* (or *variational lower bound*). For any choice of inference model $q_\phi(\boldsymbol{z}\,|\,\boldsymbol{x})$, including the choice of variational parameters ϕ, we have:

$$\log\, p_\theta(\boldsymbol{x}) = \mathbb{E}_{q_\phi(\boldsymbol{z}|\boldsymbol{x})}\left[\log\, p_\theta\,(\boldsymbol{x})\right] = \mathbb{E}_{q_\phi(\boldsymbol{z}|\boldsymbol{x})}\left[\log\, \frac{p_\theta(\boldsymbol{x}, \boldsymbol{z})}{p_\theta(\boldsymbol{z}|\boldsymbol{x})}\right]$$

$$= \mathbb{E}_{q_\phi(\boldsymbol{z}|\boldsymbol{x})}\left[\log\, \frac{p_\theta(\boldsymbol{x}, \boldsymbol{z})q_\phi(\boldsymbol{z}|\boldsymbol{x})}{q_\phi(\boldsymbol{z}|\boldsymbol{x})p_\theta(\boldsymbol{z}|\boldsymbol{x})}\right]$$

$$= \mathbb{E}_{q_\phi(\boldsymbol{z}|\boldsymbol{x})}\left[\log\, \frac{p_\theta(\boldsymbol{x}, \boldsymbol{z})}{q_\phi(\boldsymbol{z}|\boldsymbol{x})}\right] + \mathbb{E}_{q_\phi(\boldsymbol{z}|\boldsymbol{x})}\left[\log\, \frac{q_\phi(\boldsymbol{z}|\boldsymbol{x})}{p_\theta(\boldsymbol{z}|\boldsymbol{x})}\right] \tag{6.18}$$

The second term in above is the *Kullback-Leibler (KL) divergence* between $q_\phi(\boldsymbol{z}\,|\,\boldsymbol{x})$ and $p_\theta(\boldsymbol{z}\,|\,\boldsymbol{x})$, which is non-negative:

$$D_{KL}\left(q_\phi\,(\boldsymbol{z}|\boldsymbol{x})\middle\|p_\theta(\boldsymbol{z}|\boldsymbol{x})\right) \geq 0 \tag{6.19}$$

and zero if, and only if, $q_\phi(\boldsymbol{z}\,|\,\boldsymbol{x})$ equals $p_\theta\,(\boldsymbol{z}\,|\,\boldsymbol{x})$. The first term in above is the evidence lower bound (ELBO)

$$\mathcal{L}_{\theta,\phi}\,(\boldsymbol{x}) = \mathbb{E}_{q_\phi(\boldsymbol{z}|\,\boldsymbol{x})}\left[\log p_\theta\,(\boldsymbol{x}, \boldsymbol{z}) - \log q_\phi\,(\boldsymbol{z}|\,\boldsymbol{x})\right] \tag{6.20}$$

KL divergence is non-negative so the ELBO is a lower bound on the log-likelihood of the data.

$$\mathcal{L}_{\theta,\phi} = log\ p_\theta(\boldsymbol{x}) - D_{KL}\left(q_\phi(\boldsymbol{z}|\boldsymbol{x})\middle\|p_\theta(\boldsymbol{z}|\boldsymbol{x})\right) \leq log\ p_\theta(\boldsymbol{x}) \tag{6.21}$$

So, the KL divergence $D_{KL}(q_\phi(\boldsymbol{z}\,|\,\boldsymbol{x})\|p_\theta(\boldsymbol{z}\,|\,\boldsymbol{x}))$ determines two distances:

1. The KL divergence of the approximate posterior from the true posterior;
2. The gap between the ELBO and the marginal likelihood $\log p_\theta(\boldsymbol{x})$; the better $q_\phi\ (\boldsymbol{z}\,|\,\boldsymbol{x})$ approximates the $p_\theta\ (\boldsymbol{z}\,|\,\boldsymbol{x})$, the smaller the gap.

Equations above show that maximization of the ELBO w.r.t. the parameters θ and ϕ will concurrently optimize the two things we care about:

1. It will approximately maximize the marginal likelihood $p_\theta\ (\boldsymbol{x})$. This means that the generative model will become better.
2. It will minimize the KL divergence of the approximation $q_\phi\ (\boldsymbol{z}\,|\,\boldsymbol{x})$ from the true posterior $p_\theta\ (\boldsymbol{z}\,|\,\boldsymbol{x})$, so the inference model becomes better.

We can start out with random initial values of ϕ and θ, and stochastically optimize their values until convergence. Given a dataset with i.i.d. data, the ELBO objective is the sum of individual-datapoint ELBO's:

$$\mathcal{L}_{\theta,\phi}(\mathcal{D}) = \sum_{x\in\mathcal{D}} \mathcal{L}_{\theta,\phi}(\boldsymbol{x}) \tag{6.22}$$

The individual-datapoint ELBO, and its gradient $\nabla_{\theta,\phi}\,\mathcal{L}_{\theta,\phi}\,(\boldsymbol{x})$, is, in general, intractable. However, good unbiased estimators $\tilde{\nabla}_{\theta,\phi}\,\mathcal{L}_{\theta,\phi}\,(\boldsymbol{x})$ exist, such that we can still perform minibatch SGD.

Gradients of the ELBO w.r.t. the generative model parameters are:

$$\begin{aligned}\nabla_{\theta,\phi}\,\mathcal{L}_{\theta,\phi}\,(\boldsymbol{x}) &= \nabla_\theta\,\mathbb{E}_{q_\phi(\boldsymbol{z}|\,\boldsymbol{x})}\left[\log p_\theta\,(\boldsymbol{x},\boldsymbol{z}) - \log q_\phi\,(\boldsymbol{z}|\,\boldsymbol{x})\right]\\ &= \mathbb{E}_{q_\phi(\boldsymbol{z}|\,\boldsymbol{x})}\left[\nabla_\theta\left(\log p_\theta\,(\boldsymbol{x},\boldsymbol{z}) - \log q_\phi\,(\boldsymbol{z}|\,\boldsymbol{x})\right)\right]\end{aligned} \tag{6.23}$$

$$\nabla_{\theta,\phi}\,\mathcal{L}_{\theta,\phi}\,(\boldsymbol{x}) \approxeq \nabla_\theta\left(log\ p_\theta\,(\boldsymbol{x},\boldsymbol{z}) - log\ q_\phi\,(\boldsymbol{z}|\boldsymbol{x})\right) = \nabla_\phi\,(log\ p_\theta\,(\boldsymbol{x},\boldsymbol{z})) \tag{6.24}$$

The last line is a simple Monte Carlo estimator, where $\boldsymbol{z}$ is a random sample from $q_\phi\ (\boldsymbol{z}\,|\,\boldsymbol{x})$. Computing gradients w.r.t. the variational parameters, ϕ, are tricky:

$$\begin{aligned}\nabla_\phi\,\mathcal{L}_{\theta,\phi}\,(\boldsymbol{x}) &= \nabla_\phi\,\mathbb{E}_{q_\phi(\boldsymbol{z}|\,\boldsymbol{x})}\left[\log p_\theta\,(\boldsymbol{x},\boldsymbol{z}) - \log q_\phi\,(\boldsymbol{z}|\,\boldsymbol{x})\right]\\ &\neq \mathbb{E}_{q_\phi(\boldsymbol{z}|\,\boldsymbol{x})}\left[\nabla_\phi\left(\log p_\theta\,(\boldsymbol{x},\boldsymbol{z}) - \log q_\phi\,(\boldsymbol{z}|\,\boldsymbol{x})\right)\right]\end{aligned} \tag{6.25}$$

Data:
 $\mathcal{D}$: Dataset
 $q_\phi(\mathbf{z}|\mathbf{x})$: Inference model
 $p_\theta(\mathbf{x},\mathbf{z})$: Generative model
Result:
 $\boldsymbol{\theta},\boldsymbol{\phi}$: Learned parameters
$(\boldsymbol{\theta},\boldsymbol{\phi}) \leftarrow$Initialize parameters
while *SGD* not converged **do**
 $\mathcal{M} \sim \mathcal{D}$ (Random minibatch of data)
 $\epsilon \sim p(\epsilon)$ (Random noise for each datapoint in $\mathcal{M}$)
 Compute $\widetilde{\mathcal{L}}_{\theta,\phi}(\mathcal{M},\epsilon)$ and it's gradients $\nabla_{\theta,\phi}\ \widetilde{\mathcal{L}}_{\theta,\phi}\ (\mathcal{M},\epsilon)$
 Update $\boldsymbol{\theta},\boldsymbol{\phi}$ using SGD optimizer
end

Fig. 6.8 Stochastic optimization of ELBO. This is a doubly stochastic optimization procedure.

In case of continuous latent variables, we can use a *reparameterization trick* for computing unbiased estimator for $\nabla_{\theta,\phi}\,\mathcal{L}_{\theta,\phi}\,(\boldsymbol{x})$. This stochastic estimate allows us to optimize the ELBO using SGD (see Fig. 6.8).

For continuous latent variables and a differentiable encoder and generative model, the ELBO can be straightforwardly differentiated w.r.t. both ϕ and θ through a change of variables, also called the reparameterization trick (Kingma & Welling, 2013; Rezende, Mohamed, & Wierstra, 2014).

First, we express the random variable $\boldsymbol{z}\sim q_\phi\ (\boldsymbol{z}|\boldsymbol{x})$ as some differentiable (and invertible) transformation of another random variable $\boldsymbol{\epsilon}$, given $\boldsymbol{z}$ and $\boldsymbol{\phi}$:

$$\boldsymbol{z} = g(\boldsymbol{\epsilon},\, \boldsymbol{\phi},\, \boldsymbol{x}) \tag{6.26}$$

where the distribution of random variable $\boldsymbol{\epsilon}$ is independent of $\boldsymbol{x}$ or $\boldsymbol{\phi}$.

Given such a change of variable, expectations can be rewritten in terms of $\boldsymbol{\epsilon}$,

$$\mathbb{E}_{q_\phi(\boldsymbol{z}|\boldsymbol{x})}\ [f(\boldsymbol{z})] = \mathbb{E}_{p(\boldsymbol{\epsilon})}\ [f(\boldsymbol{z})] \tag{6.27}$$

where $z = g(\boldsymbol{\epsilon},\boldsymbol{\phi},\boldsymbol{x})$ and the expectation and gradient operators become commutative, and we can form a simple Monte Carlo estimator:

$$\nabla_\phi\ \mathbb{E}_{q_\phi(\boldsymbol{z}|\boldsymbol{x})}\ [f(\boldsymbol{z})] = \nabla_\phi\ \mathbb{E}_{p(\boldsymbol{\epsilon})}\ [f(\boldsymbol{z})] = \mathbb{E}_{p(\boldsymbol{\epsilon})}\ \left[\nabla_\phi f(\boldsymbol{z})\right] \approxeq \nabla_\phi f(\boldsymbol{z}) \tag{6.28}$$

where $\boldsymbol{z} = g(\boldsymbol{\phi},\boldsymbol{x},\boldsymbol{\epsilon})$ with random noise sample $\boldsymbol{\epsilon}\sim p(\boldsymbol{\epsilon})$. See Fig. 6.9 for an illustration.

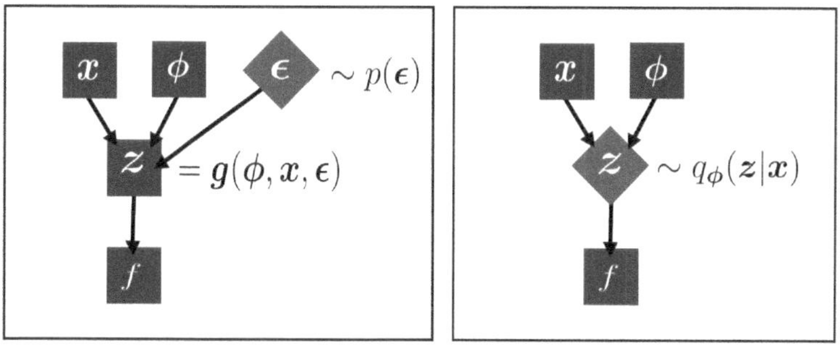

Fig. 6.9 Reparameterization trick. This change of variables helps computing gradients.

Under the reparameterization, we can replace an expectation w.r.t. $q_\phi(\boldsymbol{z}|\boldsymbol{x})$ with one w.r.t. $p(\boldsymbol{\epsilon})$. The ELBO can be rewritten as:

$$\begin{aligned}\mathcal{L}_{\theta,\phi}(\boldsymbol{x}) &= \mathbb{E}_{q_\phi(\boldsymbol{z}|\boldsymbol{x})}\left[\log p(\boldsymbol{x},\boldsymbol{z}) - \log q_\phi(\boldsymbol{z}|\boldsymbol{x})\right] \\ &= \mathbb{E}_{p(\boldsymbol{\epsilon})}\left[\log p(\boldsymbol{x},\boldsymbol{z}) - \log q_\phi(\boldsymbol{z}|\boldsymbol{x})\right]\end{aligned} \tag{6.29}$$

where $\boldsymbol{z}=g(\boldsymbol{\epsilon},\boldsymbol{\phi},\boldsymbol{x})$.

As a result, we can form a simple Monte Carlo estimator $\widetilde{\mathcal{L}}_{\theta,\phi}(\boldsymbol{x})$ of the individual-datapoint ELBO where we use a single noise sample $\boldsymbol{\epsilon}$ from $p(\boldsymbol{\epsilon})$:

$$\boldsymbol{\epsilon} \sim p(\boldsymbol{\epsilon}) \tag{6.30}$$

$$\boldsymbol{z} = g(\boldsymbol{\phi}, \boldsymbol{x}, \boldsymbol{\epsilon}) \tag{6.31}$$

$$\widetilde{\mathcal{L}}_{\theta,\phi}(\boldsymbol{x}) = \log p_\theta(\boldsymbol{x},\boldsymbol{z}) - \log q_\phi(\boldsymbol{z}|\boldsymbol{x}) \tag{6.32}$$

The resulting gradient $\nabla\widetilde{\mathcal{L}}_{\theta,\phi}(\boldsymbol{x})$ is used to optimize the ELBO using minibatch SGD. Detailed explanation, more examples, and pros and cons of VAE can be found in Kingma (2017).

2. Applications

In this section, we will present some utilizations of autoencoders in chemistry, biological and medical sciences. Of course, applications are not limited to these, and the basic idea presented in the previous section is suitable for other potential uses.

2.1 De novo molecular design

One of the most exciting fields in computational chemistry is the design and discovery of new molecules with desirable chemical and pharmaceutical

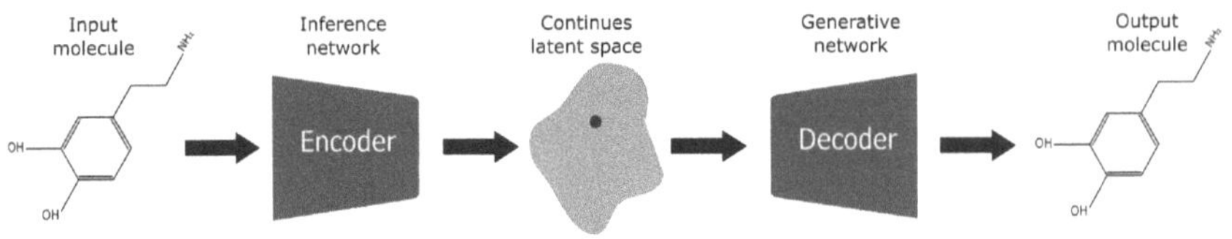

Fig. 6.10 Variational autoencoder for de novo molecular design. Inference and generative networks are used for molecular design.

properties from scratch, the so-called de novo molecular design. Variational autoencoder models can be used to explore the space of chemical molecules. The inference network is used to map the space of chemical molecules into a continuous space so that mathematical operations can be performed on objects in this space. The generative network can also be used to produce new unseen chemical molecules. The basic idea can be seen in Fig. 6.10.

Blaschke, Olivecrona, Engkvist, Bajorath, and Chen (2018) have used the variational autoencoder to develop a model for generating chemical molecules. They used three CNN layers followed by two fully connected neural layers as an encoder. The output of the last encoder layer is interpreted as the mean and variance of a Gaussian distribution. A random point is sampled using these parameters and used as input for the decoder. The decoder uses a fully connected neural layer followed by three layers of RNNs, built using GRU cells (Chung et al., 2014). The last GRU layer defines the probability distribution over all possible outputs at each position in the output sequence. In the training mode, the input of the last layer is a concatenation of the output of the previous layer and token from the target SMILES (in the training set). In contrast, in the generation mode, the input of the last layer is a concatenation of the output of the previous layer and the sampled token from the previous GRU cell. This method is called teacher forcing (Williams & Zipser, 1989) and is known to increase the performance on long text generation tasks. The network architecture is shown in Fig. 6.11.

They used Bayesian Optimization (BO) (Snoek, Larochelle, & Adams, 2012) in latent space to find molecules with desirable properties. For this purpose, they applied an objective function that estimates the probability of being active against a specific target protein. Gaussian process (Rasmussen & Williams, 2005) models with 100 starting points have been developed to predict the score of each molecule using the latent space representation. The desired points identified in the latent space are then mapped to the smiles representation using the decoder network. This process has been repeated 500 times to find the best answers.

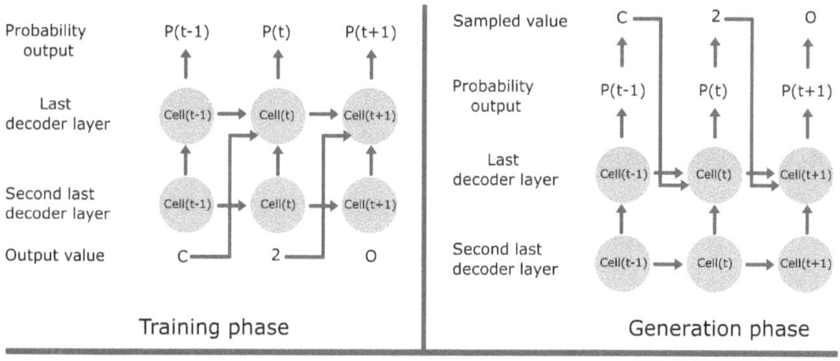

Fig. 6.11 Decoder architecture. Decoder architecture in training and generation phases.

Finally, they used their model to identify novel active molecules against dopamine receptor D2. Their results show that novel structures predicted to be active are identified by the BO search, and this indicates that AE is a valuable approach for tackling inverse QSAR problems.

Gómez-Bombarelli et al. (2018) also used an autoencoder with an encoder consisting of three layers of convolution and a decoder of three layers of GRU to build the model. A significant problem in such models is the validity of the chemical structure of the novel molecules. Jin, Barzilay, and Jaakkola (2018) trained a variational autoencoder and used a connection table to ensure the correctness of the proposed structures.

2.2 Feature extraction in virtual metrology

In semiconductor manufacturing, one of the most extensively employed data-driven applications is *virtual metrology*, where a costly or unmeasurable variable is estimated by means of cheap and easy to obtain measures that are already available in the system. Maggipinto, Masiero, Beghi, and Susto (2018) present a deep learning method for unsupervised feature extraction based on convolutional autoencoders. The proposed method is tested on a real dataset for etch rate estimation.

2.3 Chemometric data analysis

A classic problem in chemometric data analysis is the curse of dimensionality. To reduce the dimension of the data, Bilal, Ullah, and Ullah (2019) used an L2-regularized autoencoder. Also, in this study, the Gaussian process has been used to build a regressor model on hidden space. Finally, the results of the model are evaluated on orange juice and wine datasets. Usually, in such models, there is a trade-off between model complexity and generalization power. By solving a Pareto optimization problem, they have tried to select model hyperparameters in a way that maximizes generalization.

2.4 Cancer subtype classification

Hematopoietic cancer is a malignant transformation in immune system cells. Park, Batbaatar, Piao, Theera-Umpon, and Ryu (2021) used an autoencoder model to find salient features and build predictive models for hematopoietic cancer subtype classification. They have also used the oversampling method SMOTE (Chawla, Bowyer, Hall, & Kegelmeyer, 2002) to resolve the difficulty of imbalanced datasets. Another advantage of this study was that the SHAP (Lundberg & Lee, 2017) method was used to interpret the predictive model and explain primary genes and proteins for each cancer subtype. The dataset was gene expression data in 2457 patients with hematopoietic cancer, classified into five subtypes. They summed the reconstruction loss on variational autoencoder and classification loss in the cost function and used various traditional machine learning classifiers (e.g., ANN, RF, KNN, SVM). Observations show that autoencoder has a better performance in dimension reduction compared to PCA and NMF.

2.5 Dimensionality reduction in spectrometry imaging

The spectrometry method has become a dominant and efficient method in chemistry, pharmacy, and biology. New experiments in this field can generate hundreds of gigabytes of data, and as a result, their analysis requires powerful algorithms. Most methods of analyzing this data begin with a dimension reduction step. Linear methods have been used to reduce the size of this data, but the use of nonlinear methods helps reduce information loss in this process. Thomas, Race, Steven, Gilmore, and Bunch (2016) used an autoencoder to decrease the size of about 165,000 pixels and more than 7000 spectral channels to a much smaller number of essential features. In order to avoid overfitting the training data, they employ a standard L2 norm weight regularization. They applied this procedure to mouse brain data, and autoencoder has shown higher performance than PCA and NMF.

2.6 Parkinson's disease progression recognition

Parkinson's disease is a progressive nervous system disorder that affects movement. Symptoms start gradually, sometimes starting with a barely noticeable tremor in just one hand. Tremors are common, but the disorder also commonly causes stiffness or slowing of movement.

The Unified Parkinson's Disease Rating Scale (UPDRS) has four parts. Each part has multiple points that are individually scored, using zero for normal or no problems, 1 for minimal problems, 2 for mild problems, 3 for moderate problems, and 4 for severe problems. These scores are tallied to indicate the severity of the disease, with 199 points being the worst and total disability and 0 meaning no disability.

Kazak (2019) used a dataset containing 5875 voice recordings from 42 patients diagnosed with early-stage Parkinson's disease: 28 men and 14 women (about 200 recordings per patient). Each observation is described with the following variables: patient number, age, gender, time interval from baseline recruitment date, 16 vocal attributes based on biomedical voice measures, and Unified Parkinson's Disease Rating Scale (UPDRS) scores, used for tracking symptom progression. They trained a stacked autoencoder for predicting UPDRS scores. The results show that this method is more effective than traditional linear methods.

References

Bengio, Y., Ducharme, R., Vincent, P., & Janvin, C. (2003). A neural probabilistic language model. *The Journal of Machine Learning Research*, *3*, 1137–1155.

Bilal, M., Ullah, M., & Ullah, H. (2019). Chemometric data analysis with autoencoder neural network. *Electronic Imaging*, *2019*(1), 671–679.

Blaschke, T., Olivecrona, M., Engkvist, O., Bajorath, J., & Chen, H. (2018). Application of generative autoencoder in de novo molecular design. *Molecular Informatics*, 1700123.

Chawla, N. V., Bowyer, K. W., Hall, L. O., & Kegelmeyer, W. P. (2002). SMOTE: Synthetic minority over-sampling technique. *Journal of Artificial Intelligence Research*, *16*, 321–357.

Cho, K., Van Merriënboer, B., Gulcehre, C., Bahdanau, D., Bougares, F., Schwenk, H., et al. (2014). Learning phrase representations using RNN encoder-decoder for statistical machine translation. *arXiv Preprint arXiv:1406.1078*.

Chung, J., Gulcehre, C., Cho, K., & Bengio, Y. (2014). Empirical evaluation of gated recurrent neural networks on sequence modeling. *arXiv Preprint arXiv:1412.3555*.

Fukushima, K., & Miyake, S. (1982). Neocognitron: A self-organizing neural network model for a mechanism of visual pattern recognition. *Competition and Cooperation in Neural Nets*, 267–285.

Gómez-Bombarelli, R., Wei, J., Duvenaud, D., Hernández-Lobato, J., Sánchez-Lengeling, B., Sheberla, D., et al. (2018). Automatic chemical design using a data-driven continuous representation of molecules. *ACS Central Science*, 268–276.

Goodfellow, I., Bengio, Y., & Courville, A. (2016). *Deep learning*. MIT Press.

Hinton, G. E., & Salakhutdinov, R. R. (2006). Reducing the dimensionality of data with neural networks. *Science*, *313*(5786), 504–507.

Hochreiter, S., & Schmidhuber, J. (1997). Long short-term memory. *Neural Computation*, *9*(8), 1735–1780.

Jin, W., Barzilay, R., & Jaakkola, T. (2018). Junction tree variational autoencoder for molecular graph generation. *International Conference on Machine Learning*, 2323–2332.

Kazak, V. (2019). *Unsupervised feature extraction with autoencoder: For the representation of Parkinson's disease patients*. Universidade NOVA de Lisboa.

Kingma, D. (2017). *Variational inference & deep learning: A new synthesis.* Intelligent Sensory Information Systems.
Kingma, D., & Welling, M. (2013). Auto-encoding variational bayes. *arXiv Preprint arXiv:1312.6114.*
Kohlbrenner, M., Hofmann, R., Ahmmed, S., & Kashef, Y. (2017). *Pre-training CNNs using convolutional autoencoders.* https://pdfs.semanticscholar.org/e1ab/3b9dee2da2007 8464f4ad8deb523b5b1792e.pdf.
Lecun, Y. (1987). *PhD thesis: Modeles connexionnistes de l'apprentissage (connectionist learning models).* NYU.
LeCun, Y., Boser, B., Denker, J. S., Henderson, D., Howard, R. E., Hubbard, W., et al. (1989). Backpropagation applied to handwritten zip code recognition. *Neural Computation, 1*(4), 541–551.
LeCun, Y., Bottou, L., Bengio, Y., & Haffner, P. (1998). Gradient-based learning applied to document recognition. *Proceedings of the IEEE, 86*(11), 2278–2324.
Lundberg, S. M., & Lee, S.-I. (2017). A unified approach to interpreting model predictions. In *Proceedings of the 31st international conference on neural information processing systems* (pp. 4768–4777).
Maggipinto, M., Masiero, C., Beghi, A., & Susto, G. (2018). A convolutional autoencoder approach for feature extraction in virtual metrology. *Procedia Manufacturing,* 126–133.
Makhzani, A., & Frey, B. (2014). A winner-take-all method for training sparse convolutional autoencoders. In *NIPS deep learning workshop.*
Ng, A., et al. (2011). Sparse autoencoder. *CS294A Lecture Notes, 72,* 1–19.
Olshausen, B. A., & Field, D. J. (1997). Sparse coding with an overcomplete basis set: A strategy employed by V1? *Vision Research, 37*(23), 3311–3325.
Park, K. H., Batbaatar, E., Piao, Y., Theera-Umpon, N., & Ryu, K. H. (2021). Deep learning feature extraction approach for hematopoietic cancer subtype classification. *International Journal of Environmental Research and Public Health, 18*(4), 2197.
Rasmussen, C., & Williams, C. (2005). *Gaussian processes for machine learning (adaptive computation and machine learning)* (pp. 69–106). Cambridge, MA: The MIT Press.
Rezende, D., Mohamed, S., & Wierstra, D. (2014). Stochastic backpropagation and approximate inference in deep generative models. *International Conference on Machine Learning,* 1278–1286.
Schuster, M., & Paliwal, K. K. (1997). Bidirectional recurrent neural networks. *IEEE Transactions on Signal Processing, 45*(11), 2673–2681.
Snoek, J., Larochelle, H., & Adams, R. (2012). Practical Bayesian optimization of machine learning algorithms. *Advances in Neural Information Processing Systems.*
Socher, R., Huang, E., Pennin, J., Manning, C. D., & Ng, A. (2011). Dynamic pooling and unfolding recursive autoencoders for paraphrase detection. *Advances in Neural Information Processing Systems, 24.*
Soulie, F. F., Gallinari, P., Le Cun, Y., & Thiria, S. (1987). *Evaluation of network architectures on test learning tasks.* NYU.
Sutskever, I., Vinyals, O., & Le, Q. V. (2014). Sequence to sequence learning with neural networks. *Advances in Neural Information Processing Systems,* 3104–3112.
Thomas, S. A., Race, A. M., Steven, R. T., Gilmore, I. S., & Bunch, J. (2016). Dimensionality reduction of mass spectrometry imaging data using autoencoders. In *2016 IEEE symposium series on computational intelligence (SSCI)* (pp. 1–7).
Vincent, P., Larochelle, H., Lajoie, I., Bengio, Y., Manzagol, P.-A., & Bottou, L. (2010). Stacked denoising autoencoders: Learning useful representations in a deep network with a local denoising criterion. *Journal of Machine Learning Research, 11*(12).
Williams, R., & Zipser, D. (1989). A learning algorithm for continually running fully recurrent neural networks. *Neural Computation,* 270–280.

CHAPTER 7

Uniqueness in resolving multivariate chemical data

Hamid Abdollahi[a], Somaiyeh Khodadadi Karimvand[b], and Somaye Vali Zade[c]
[a]Department of Chemistry, Institute for Advanced Studies in Basic Sciences (IASBS), Zanjan, Iran
[b]Department of Chemistry, Institute for Advanced Studies in Basic Sciences, Zanjan, Iran
[c]Halal Research Center of IRI, Food and Drug Administration, Ministry of Health and Medical Education, Tehran, Iran

1. Introduction

Self-modeling curve resolution (SMCR) techniques are widely applied for resolving chemical data to the pure-component spectra and composition profiles. In most cases, the resolution of bilinear chemical data sets using self-modeling curve resolution (SMCR) methods produces a range of feasible solutions in multicomponent systems. Hence, the most significant challenge of the chemometricians who use the SMCR methods is to reduce the range of feasible solutions by applying proper information in the form of constraints. The mathematical solutions generated by SMCR obey the applied constraints coming from a priori physicochemical information about the system under investigation.

The appropriate constraints can narrow down the band of feasible solutions, even in most favorable cases, can produce unique solutions, and can improve the accuracy of the results of SMCR methods. This highlights the importance of studying the constraints used in SMCR methods, and the need for new constraints to achieve more accurate results. Several studies have demonstrated that applying some constraints in SMCR methods, including trilinearity, equality, zero concentration region, correspondence, local rank, and non-negativity, can result in unique profiles.

Undoubtedly, the non-negativity constraint is a natural and most frequently used constraint for resolving many kinds of chemical data, especially spectroscopic data that significantly restrict the feasible solutions (Lawton & Sylvestre, 1971). Interestingly, under some specific conditions, this simple natural constraint is sufficient for achieving a unique profile (partial uniqueness) (Rajkó, Abdollahi, Beyramysoltan, & Omidikia, 2015). In 2015, the

Machine Learning and Pattern Recognition Methods in Chemistry from Multivariate and Data Driven Modeling
https://doi.org/10.1016/B978-0-323-90408-7.00006-X

necessary conditions for a partially unique resolution based on the non-negativity constraint were proposed with the so-called data-based uniqueness (DBU) theorem. DBU can predict the unique profiles when imposing only the non-negativity constraint by investigating the selectivity and zero values of the resolved profiles.

Manne (1995) proposed "resolution theorems" that describe the conditions for getting unique concentration and spectral profiles under local-rank information (Manne, 1995). However, successful prediction of the unique profiles based on the resolution theorems is subject to the implementation of the zero-region constraints in iterative curve resolution methods (Tavakkoli, Rajkó, & Abdollahi, 2018). Trilinearity constraint is another powerful constraint for exploiting the unique trilinear profile during a multiset data analysis (Ghaffari & Abdollahi, 2018). Trilinearity constraint can be imposed when the experiments are repeated under different initial conditions. It follows a trilinear model where individual pure components can be expressed as the product of a triad of linear vectors. In addition to the mentioned constraints, entering the information about the shape of a specific profile to the SMCR methods as the equality constraint leads to a unique solution (Beyramysoltan, Rajkó, & Abdollahi, 2013).

The effect of any available information was also investigated on the chemical reaction system such as monotonicity and unimodality constraints to reduce ambiguity (Zade, Sawall, Neymeyr, & Abdollahi, 2019). In addition to these strong real physical/chemical constraints often resulting in a unique solution, there are some other mathematical constraints where the unique solutions are not necessarily the true ones (Khodadadi Karimvand & Abdollahi, 2018). Frequently used local-rank constants with a large impact on the range of feasible solutions of SMCR methods are based on information derived from the mathematical properties of the data. According to this constraint, the number of chemical components in local windows of the data is assumed to be the same as the windows rank. This constraint also creates a limited range of solutions by applying the zero region for the missing profile.

The other important group of constraints that is used for narrowing the feasible band is information about the absence of a chemical species in a part of the contribution (or spectral) profile. This information can be applied as zero-region constraint when analyzing a single data matrix in iterative curve resolution methods. In case of multiset data analysis, the information about the presence/absence of each component in the matrices analyzing together is considered as the correspondence of species. We refer to this group of constraint as "nonexistence information." The circumstances that the non-existence information can result in a unique solution are derived and can be find in Akbari Lakeh, Rajkó, and Abdollahi (2017).

Defining and incorporating hard modeling constraint generally reduces the range of feasible solutions and often results in unique solutions. The hard constraint is introduced to force some or all of the concentration profiles to follow chemical model, which is refined at each iterative cycle of the optimization process of soft modeling. Therefore, the concentration and spectral profiles, and the physicochemical parameters (e.g., kinetic rate constants) can be obtained even though there are compounds or interferences that do not take part in the explicit model.

The mathematical concept of duality was initially introduced by Henry in the chemometrics literature (Henry, 2005). Later, Rajkó showed that it is possible to determine the feasible regions of two-component systems using the duality principles. Duality as a mathematical concept provides a relation between the row and column spaces of a data matrix. Simply, the duality principle states that each data point in a space (e.g., the column space) corresponds to one special directed hyperplane in the dual space (the row space).

In this chapter, the circumstances and the rules that govern to achieve the unique real solutions through any of the constraint will be explained. A collected knowledge in this subject is completely rare in literature.

2. Non-negativity constraint

Non-negativity is the first and most general natural constraint in different fields of chemistry. For instance, when the data are measurements of counts, color intensities, amount of absorption, or spectral amplitudes, negative numbers do not have a physical interpretation. Thus, non-negativity is frequently applied as the minimal constraint of the multivariate curve resolution (MCR) studies. Under non-negativity constraints, feasible regions are bounded (Lawton & Sylvestre, 1971). The non-negativity constraint reduces the range of feasible solutions of a curve resolution problem. In some favorable cases, a unique solution can be obtained by applying only the non-negativity constraints. Data-based uniqueness (DBU) refers to decompositions that non-negativity constraint is sufficient to get a unique solution (Rajkó et al., 2015). Different levels of uniqueness were defined in Rajkó et al. (2015) as non-uniqueness, fractional uniqueness, partial uniqueness, and full uniqueness.

2.1 Data-based uniqueness

Data-based uniqueness condition can be defined as the following theorem in two equivalent forms in general, i.e., the theorem is formulated in such a way that is independent of the number of the components.

2.1.1 *Data-based uniqueness theorem (Rajkó et al., 2015)*

Form 1: If the Borgen plot (or its generalization for arbitrary dimensions) contains a point belonging to the coincidence of vertices of the inner and outer polygon (inner and outer polyhedrons), and, as a result, this point will be the matching vertices of the two Borgen triangles (all Borgen simplices), then the point signifies the unique solution for the component.

Form 2: If a component has a selective window in one direction/way/mode and considering the other direction/way/mode, this component has a sub-window in which the contribution of this component is zero while all interfering compounds appear inside this window, then the component can be uniquely decomposed in this second direction/way/mode.

2.2 Borgen plot configuration in case of unique resolution of profile(s)

For ordinary (general) cases of spectroscopy, the non-negativity condition for profiles on its own is not enough for a unique resolution of the profiles except for some special cases. In this section, some hypothetical spectroscopic (or any two-way) data sets are represented for conditions, which will lead to uniquely resolved profiles under non-negativity constraint. Figs. 7.1 and 7.2 depict the examples of Borgen plot patterns for data sets where data-based uniqueness occurs. Specific property of these geometries is the

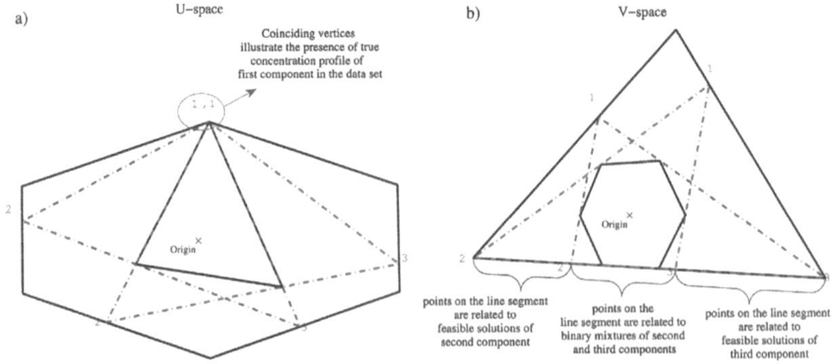

Fig. 7.1 An illustration of abstract space of data set when a unique profile occurred. (A) U-space (abstract concentration space). (B) V-space (abstract spectral space).

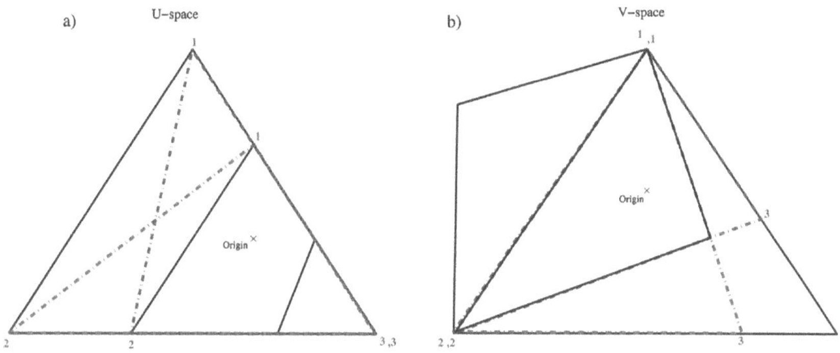

Fig. 7.2 A depiction of abstract space of a data set when three unique profiles observable: two unique profiles for two components in V-space and one unique profile for complementary component in dual space (U-space). (A) U-space (abstract concentration space). (B) V-space (abstract spectral space).

coincidence of one vertex of the outer polygon with one vertex of the inner polygon. The convex geometry of Borgen plot ensures that this phenomenon causes the coincidence of one vertex of one Borgen triangle with the vertex of the other Borgen triangle as it can be seen in U-spaces in Figs. 7.1A and 7.2A and V-space in Fig. 7.2B. Therefore, the permitted area will be a single point, which defines the feasible region as point accordingly; this point can be transformed to the unique profile. There will be (partially or entirely) overriding sides between Borgen triangles in the dual space on the side (or with another term: facet) of the outer polygon corresponding to the intersection point.

These partially or entirely overriding common sides can be interpretable as duality result of the intersection point in complementary abstract space; based on the duality concept (Rajkó, 2006), the coordinates of one profile, i.e., a point, define a line in its dual space. This line (for example, one is shown in Fig. 7.1B) consists of three segments: the two end-fragments are the outcome of the restricted feasible regions of complementary components associated with a component that has a unique profile (Beyramysoltan et al., 2013). Points belonging to the middle line segment are associated with binary mixtures that are linear combination, according to the convexity, of pure profiles of complementary components.

This geometry, i.e., touch between vertices of Borgen triangles in one point, is of particular concern for indicating the uniqueness. Therefore, uniqueness can be detected by the stated structural information of the system, which is deduced from Borgen plot. A new term for this type of uniqueness is coined, i.e., data-based uniqueness, since the analyst does not need to know anything about the properties of the true profiles. On the contrary, Manne's theorems (Manne, 1995) define the profile-based uniqueness, because we should prescribe some properties of the true profiles in advance for that type of uniqueness.

3. Trilinearity constraint

Trilinearity constraint, which is based on the chemical information, guarantees the accuracy of unique profiles in the most favorite cases. Due to the trilinearity constraint, data arrays should contain some similar factors in their slices/modes, while the factors should be independent from experimental conditions. Thus, factors change proportionally (Cattell, 1944). Trilinearity is a very strong constraint that enforces trilinear decompositions to have unique solutions under mild conditions (Alier, Felipe, Hernández, & Tauler, 2011).

In the most second-order data sets with trilinear structure, it is possible to extract the trilinear profile of the given analyte accurately. In this case, the trilinear behavior of the specific profile is the available real information, and knowing this information depends on the primary knowledge of the studied system. Applying this information in the qualified three-way data resulted in a unique solution for the profile of the component of interest.

Matrix-augmented-multivariate curve resolution-alternating least square, MA-MCR-ALS, as an extension of MCR-ALS, can be easily adapted to the trilinear analysis of the three-way data sets generated from data matrices with the same row or column and taken the advantages of their structure. When trilinearity constraint is incorporated into the MA-MCR-ALS as a constraint, profiles of a component in different data matrices are forced to have common shape during each ALS optimization and they may differ by a scaling factor. When this constraint is inserted into the ALS iterative optimization procedure, it forces the shape of the loading vectors to be the same in all submatrices. A wonderful aspect of the trilinearity constraint in the ALS is that the application of this constraint for each component can be optional.

Accuracy of the measurement is the main goal of analytical chemists. Programmers and mathematicians have been developing algorithms, which may produce "unique solution" based on the available information. Most of the algorithms provide only a solution at the end of the procedure but it might not be necessarily the true solution (Rajkó, 2010). Trilinearity constraint, which is based on the chemical information, guarantees the accuracy of the resolved profiles due to the uniqueness property. Trilinearity may be incorporated in the algorithm such as MA-MCR-ALS or can be an intrinsic property of a model like PARAllel FACtor analysis (PARAFAC), rank annihilation factor analysis (RAFA), generalized rank annihilation method (GRAM), direct trilinear decomposition (DTLD), etc. (Abdollahi & Nazari, 2003; Sanchez & Kowalski, 1986).

Let's consider $\mathbf{A}_{I,F}$, $\mathbf{B}_{J,F}$, and $\mathbf{C}_{K,F}$ component matrices of a three-way array and k stands for Kruskal's rank of the components matrices, and F is an array rank. According to the Harshman's, it is sufficient for uniqueness to have $\mathbf{A}_{I,F}$ and $\mathbf{B}_{J,F}$ of full-column rank, and $\mathbf{C}_{K,F}$ of k-rank 2 or higher (Harshman, 1970). Kruskal has proved that uniqueness is definite when:

$$K_A + K_B + K_C \geq 2F + 2$$

Despite the Harshman sufficient condition, Kruskal has illustrated that a PARAFAC solution may be unique even if none of the component matrices have a full-column rank.

When $F=1$, the Kruskal condition cannot be achieved. However, Harshman has already proven uniqueness condition for the mentioned case (Harshman, 1970). An intuitive and accessible proof of Kruskal condition is provided by Stegeman and Sidiropoulos (2007). Sidiropoulos and Bro generalized Kruskal fundamental result about the uniqueness of the trilinear decomposition (triadic decomposition) of the three-way array to the cases with multilinear decomposition of higher-order arrays (Sidiropoulos & Bro, 2000). Sajjadi et al. have visualized the rotational ambiguity in the PARAFAC model for the two-component rank overlapped systems (Abdollahi & Sajjadi, 2010). Omidikia et al. have investigated the uniqueness property of different three-component systems by direct visualization of the feasible bands.

3.1 Theoretical basis (Ghaffari & Abdollahi, 2018)

3.1.1 Duality

Duality concept represents the relation between column and row spaces of the bilinear data matrices due to the bilinear structure. Based on the Bouguer-Lambert-Beer law, any bilinear data matrix can be decomposed as follows:

$$\mathbf{D} = \mathbf{C}\,\mathbf{A}^{T} + \mathbf{E} \tag{7.1}$$

where $\mathbf{D}$ is a measured data matrix, $\mathbf{E}$ contains the residual parts of the data that has not been modeled by a decomposition procedure, and $\mathbf{C}$ and $\mathbf{A}$ contains concentration and spectral profiles, respectively. Moreover, singular value decomposition (SVD) can be employed to define the basis of the row (V) and column spaces (U) of a data matrix:

$$\mathbf{D} = \mathbf{US}\,\mathbf{V}^{\mathrm{T}} + \mathbf{E}' = \mathbf{X}\,\mathbf{V}^{\mathrm{T}} + \mathbf{E}' = \mathbf{U}\,\mathbf{Y}^{\mathrm{T}} + \mathbf{E}' \tag{7.2}$$

where X and Y are the coordinates of rows and columns of the **D** in its row and column space. **E**′ is the residuals or non-modeled parts of the **D**. In general, due to the experimental noise and different ways of matrix decompositions (see Eqs. 7.1 and 7.2), **E** are almost different.

Based on the duality concept, each data point in one space (which can be a column, a row, or any arbitrary point in the space) determines the coefficients of a hyperplane in its dual space. Let's suppose **x** as the coordinates of each row of the **D** in the V-space, while its corresponding dual hyperplane in the U-space can be defined as:

$$\mathbf{x}\,\mathbf{S}^{-1}\,\mathbf{z} = 0 \tag{7.3}$$

This equation demonstrates an equality relation to determine a hyperplane as the dual of the **x** in the row space. In addition, **z** is a general variable vector representing the coordinates of the points on the hyperplane in the U-space.

Furthermore, y is the coordinates of the **D** columns in the U-space, and its dual hyperplane in the V-space is:

$$\mathbf{y}\,\mathbf{S}^{-1}\,\mathbf{q} = 0 \tag{7.4}$$

where **q** is a general variable representing the coordinates of the points on the hyperplanes in the V-space. It is worth to highlight that the duality relation is not limited to a data rows and columns. All points in the U-space of a bilinear data (regardless of the position of the point) have dual hyperplanes in its V-space, and vice versa. Accordingly, a hypothetical point, h, in the U-space has a dual hyperplane in the V-space:

$$\mathbf{h}\,\mathbf{S}^{-1}\,\mathbf{q} = 0 \tag{7.5}$$

where **h** is a vector containing the coordinates of any hypothetical point in the U-space. Thus, it is possible to define its dual hyperplane based on the duality concept in the V-space.

Duality-based explanation of the trilinearity constraint

Column-wise augmented matrix $\mathbf{D}_{\mathrm{aug}}$ can be written as $[\mathbf{D}_1; \mathbf{D}_2; \mathbf{D}_3; \ldots \mathbf{D}_k]$ with the size/dimension of $\mathrm{I} \times \mathrm{J}$, where the semicolon ";" notation, which indicates the different data matrices (K matrices), are column-wise appended. In this data arrangement strategy, different data matrices are supposed to have common row vector spaces.

$$[\mathbf{D}_1;\mathbf{D}_2;\mathbf{D}_3;\ldots;\mathbf{D}_k] = [\mathbf{C}_1;\mathbf{C}_2;\mathbf{C}_3;\ldots;\mathbf{C}_k]\,\mathbf{A}^T + [\mathbf{E}_1;\mathbf{E}_2;\mathbf{E}_3;\ldots;\mathbf{E}_k]$$
$$= \mathbf{C}_{aug}\,A^T + \mathbf{E}_{aug} \quad (7.6)$$

A column-wise augmented concentration matrix, i.e., $[\mathbf{C}_1; \mathbf{C}_2; \mathbf{C}_3; \ldots \mathbf{C}_k]$, is obtained to describe the concentration changes in the different individual data matrices. $\mathbf{A}^T$ contains spectral profiles of all constituents in all sub-matrices and $\mathbf{E}_{aug}$ is the residual matrix. Eq. (7.6) shows the augmentation of K matrices with the same size of $I \times J$. $[\mathbf{D}_1; \mathbf{D}_2; \mathbf{D}_3; \ldots; \mathbf{D}_K] = \mathbf{D}_{aug} = \mathbf{C}_{aug}\,\mathbf{A}^T$. Hence, $\mathbf{D}_{aug}$ has IK rows and J columns. Considering the fact that $\mathbf{C}$, $\mathbf{A}$, and $\mathbf{C}_t$ are the component matrices related to the three modes, each of $\mathbf{D}_k$ matrices can be formulated as:

$$\mathbf{D}_k = \mathbf{C}_{diag}(\mathbf{C}_t(k,\ :))\mathbf{A}^T, \quad k = 1,\ldots,k. \quad (7.7)$$

The outer product of the respective component matrices, $\mathbf{C}$ and $\mathbf{A}^T$ in the column and row space of $\mathbf{D}_k$ with proper intensities, will generate each $\mathbf{D}_k$ matrix. The $\mathbf{C}_t$ $(k,:)$ matrix contains total concentration values of each component in the kth data set. Equation is a well-known representation of a series of matrices with the trilinearity properties. The abstract space of the $\mathbf{D}_{aug}$ can be estimated by the application of the SVD to the data matrix (r is the number of abstract mathematical factors in the $\mathbf{D}_{aug}$):

$$\mathbf{D}_{aug} = \mathbf{U}_{aug}\,\mathbf{S}_{aug}\,\mathbf{V}^T_{aug} + \mathbf{E}'_{aug} = \mathbf{X}_{aug}\,\mathbf{V}^T_{aug} + \mathbf{E}'_{aug} = \mathbf{U}_{aug}\,\mathbf{Y}^T_{aug} + \mathbf{E}'_{aug} \quad (7.8)$$

If $\mathbf{E}_{aug}$ is almost the same as the $\mathbf{E}'_{aug}$, the subspace spanned by the truncated $\mathbf{U}_{aug}$ is a good estimation of the subspace spanned by the $\mathbf{C}_{aug}$ ($\mathbf{C}_{aug} = \mathbf{U}^T_{aug}$); likewise, the subspace spanned by the truncated matrix is a good estimation of the subspace spanned by the $\mathbf{A}^T$ ($\mathbf{A} = \mathbf{T}^{-1} \times \mathbf{V}_{aug}$). $\mathbf{T}$ and $\mathbf{T}^{-1}$, as the transformation matrixes, generate $\mathbf{C}_{aug}$ and $\mathbf{A}^T$ by the linear combination of the $\mathbf{U}_{aug}$ and $\mathbf{V}_{aug}$, respectively.

As a result of the bilinear structure of the data, $\mathbf{V}_{aug}$ and $\mathbf{U}_{aug}$ spaces have a mutual duality relationship. Thus, each point in the column space of a data has a dual hyperplane in its row space, and vice versa. This mathematical relation provides an efficient tool to transfer the information of one space to its dual space. Consequently, in case of unique solution or uniqueness condition, the duality concept is a powerful approach to be studied and visualized.

When trilinearity is inserted into the ALS iterative optimization procedure, profiles of the same component in the different parts of $\mathbf{C}_{\text{aug}}$ matrix are forced to have the same shape in each iteration of the ALS and only differ with a scaling factor. Trilinearity can be implemented as follows:

(1) Selecting a column of $\mathbf{C}_{\text{aug}}$, ($\mathbf{c}^{f}{}_{\text{aug}}$, f=1... F).
(2) $\mathbf{c}^{f}{}_{\text{aug}}$ with the size of IK × 1 was reshaped to a matrix with the size of I × K.

$$\mathbf{C}^{f}{}_{\text{Mat}} = \text{reshape}\left(\mathbf{c}^{f}{}_{\text{aug}}\right),$$

where $\mathbf{C}^{f}{}_{\text{Mat}}$ is a matricized form of $\mathbf{c}^{f}_{\text{aug}}$ and reshape is the MATLAB command.

(3) SVD was applied on $\mathbf{C}^{f}{}_{\text{Mat}}$.

$$[\mathbf{u}\ \mathbf{s}\ \mathbf{v}] = \text{svd}\left(\mathbf{C}^{f}{}_{\text{Mat}}\right), f = 1\ldots\text{F}.$$

(4) Using the first principal component, $\hat{\text{C}}^{f}{}_{\text{Mat}}$ is reconstructed.

$$\hat{\mathbf{C}}^{f}{}_{\text{Mat}} = u_1\ s_1\ v^{T}{}_1.$$

(5) $\hat{\text{C}}^{f}{}_{\text{Mat}}$ was unfold to a vector with the size of IK × 1.

The aforementioned steps ensure the idea of parallel proportional profiles.

These steps are necessary to explain the uniqueness property of trilinearity constraint by the duality concept:

First, SVD should be employed for defining the basis vectors of row and column space of $\mathbf{D}_{\text{K}}$:

$$\mathbf{D}_k = \mathbf{U}_k\ \mathbf{S}_k\ \mathbf{V}_k^{\text{T}} + \mathbf{E}_k = \mathbf{X}_k\ \mathbf{V}_k^{\text{T}} + \mathbf{E}_k = \mathbf{U}_k\ \mathbf{Y}_k^{\text{T}} + \mathbf{E}_k, \quad k = 1,\ldots,K \tag{7.9}$$

In this section, $\mathbf{t}_1$ is a vector containing the coordinates of a point in the column space of $\mathbf{D}_1$, whose correspondence profile belongs to the column space of all sub-matrices of $\mathbf{D}_{\text{aug}}$, and $\mathbf{c}_1$ is a profile correspondence to $\mathbf{t}_1$ ($\mathbf{c}_1 = \mathbf{U}_1\ \mathbf{t}_1$), which belongs to all sub-matrices. Hence, based on the nonnegativity constraint, it is possible to define the dual hyperplane of the $\mathbf{t}_1$ in the V-space of the $\mathbf{D}_1$ by the following equation ($\mathbf{z}_1$ is the general variable vector representing the coordinates of an arbitrary point in the dual hyperplane):

$$\mathbf{t}_1\ \mathbf{S}_1^{-1}\ \mathbf{z}_1 = 0 \tag{7.10}$$

In the next step, the coordination of $\mathbf{c}_1$, $\mathbf{t}_i$, in the column space of the other sub-matrices must be calculated:

$$\mathbf{t}_i = \mathbf{c}_1\mathbf{U}_i, \quad i = 2,3,..,k \tag{7.11}$$

Afterward, the complementary subspace of $\mathbf{c}_1$ in all sub-matrices can be calculated based on the duality concept:

$$\mathbf{t}_i\,\mathbf{S}_i^{-1}\,\mathbf{z}_i = 0 \tag{7.12}$$

The $\mathbf{z}_i$ represents the coordinates of an arbitrary point on the defined hyperplane in the V-space of $\mathbf{D}_i$. There is K hyperplane for K sub-matrices of the augmented data matrix, and their intersection defines a total hyperplane in the row space of the $\mathbf{D}_{aug}$. The dual point of the total hyperplane in the column space of the $\mathbf{D}_{aug}$ can generate the trilinear profile, while the selected initial $\mathbf{t}_1$ is correct. Thus, all hyperplanes in the row spaces of all sub-matrices must be transformed to the row space of the $\mathbf{D}_{aug}$ by linear transformation of matrix spaces. $\mathbf{T}_k$ is a transformation matrix:

$$\mathbf{T}_k = \mathbf{V}_k^{-1}\,\mathbf{V}_{aug}, \quad k = 1,2,\ldots,K \tag{7.13}$$

Based on the linear algebra, coordinate-wise representations of vectors of a basis set can be transformed to their equivalent representations of another basis set. This transformation is called change of the basis. Therefore, it is necessary to transform r_K^{-1} arbitrary points of each hyperplane in $\mathbf{V}_K$ to the row space of the $\mathbf{D}_{aug}$ ($\mathbf{V}_{aug}$) by the transformation matrices (r_K is chemical rank of Kth sub-matrix):

$$\mathbf{Z}'_K = \mathbf{Z}_K\,\mathbf{T}_K \tag{7.14}$$

$\mathbf{Z}_k$ and $\mathbf{Z}'_K$ are the coordination of arbitrary points in the $\mathbf{V}_K$ and $\mathbf{V}_{aug}$ before and after transformation, respectively. Thereafter, the total hyperplane (coefficients of the polynomial) should be calculated by $\sum_{K=1}^{K}(r_K - 1)$ points ($\mathbf{Z}'_K$).

As a simple example, $1x+2y+3z=0$ is a hyperplane with three coefficients. Thus, [1 2 3] are the coefficients of a plane (intercept is zero) as well as coordinates of a vector which is orthogonal to the plane (intercept is zero). Accordingly, in order to find the coefficients of the plane, it is necessary to look for a vector which is orthogonal to the plane. $[x\ y\ z]$ is the coordinates of any arbitrary point on the plane. Euclidean inner product of two vectors, $(\vec{m}, \vec{w})$, can be defined as: $(\vec{m}, \vec{w}) = \overrightarrow{m^T}, \vec{w}$. Hence, $\{\overrightarrow{m^T} = [1\ 2\ 3]$, with $\vec{w} = [x;\ y;\ z]$ and $(\vec{m}, \vec{w}) = 1x+2y+3z=0\}$.

Regardless of the dimension of the hyperplane, *Null* operator in MATLAB can be implemented to calculate the coordinates of the hyperplane.

Hence, $\mathbf{t}_{aug}$ contains coefficients of the total hyperplane of all complementary components in the row space of the augmented data, and $\mathbf{Z}_{aug}$ is the coordination of any arbitrary point on a hyperplane:

$$\mathbf{t}_{aug}\ \mathbf{z}_{aug} = 0 \tag{7.15}$$

Based on the duality concept, this hyperplane sets a point, $\mathbf{t}_{aug}\ \mathbf{S}_{aug}$, in the column space of the augmented data matrix. As a consequence, linear combination of the columns of $\mathbf{U}_{aug}$ with coordination of the point ($\mathbf{t}_{aug}\ \mathbf{S}_{aug}$), can generate $\mathbf{C}_{aug}$.

$$\mathbf{C}_{aug} = \mathbf{t}_{aug}\ \mathbf{S}_{aug}\ \mathbf{U}_{aug} \tag{7.16}$$

By converting this column vector with k parts to a matrix with k columns, $\mathbf{C}_{aug}$ can be calculated:

$$\mathbf{C}_{aug} = \left[\mathbf{c}_{aug,1},\ \mathbf{c}_{aug,2},\ \ldots,\ \mathbf{c}_{aug,K}\right] \tag{7.17}$$

Due to the trilinear structure of $\mathbf{C}_{aug}$, second eigen value of the metricized $\mathbf{C}_{aug}$ should be zero. Non-zero value for the second singular value shows the non-trilinearity of the $\mathbf{C}_{aug}$. In summary, the calculated c_1 (based on t_1) is projected to the column space of all sub-matrices and total complementary subspace is calculated based on these projections. Trilinearity constraint, which works based on the proportionality property, can produce the unique solution. If c_1 belongs to the column space of all sub-matrices, the second eigen value of $\mathbf{C}_{aug}$ should be close to zero; otherwise, c_1 does not belong to all sub-matrices.

4. Nonexistence information

Information on the absence of a chemical species in a part of the presence region of components is introduced by several constraints in literature, and this information is used to reduce the range of feasible solutions. In multiset data analysis, chemical information related to the presence/absence of each component in the matrices analyzing together is considered as the correspondence of species (Abdollahi & Sajjadi, 2010). It can be said that applying this constraint is, in fact, equivalent to applying a zero concentration region for species not present in each data matrix (Tavakkoli et al., 2018).

With correct a priori information on the zero concentration regions of components, Manne's theorems will be successfully able to produce the unique solution(s) provided that the necessary conditions were fulfilled. However, these conditions regard the properties of the mostly unknown true profiles of the components. That is why, we can call the resulted phenomenon as the profile-based uniqueness instead of using the term "data-based uniqueness" (Rajkó et al., 2015).

In the literature of chemometrics, resolution theorems of Manne are discussed and explained by local-rank information. But recently, it was shown that the local rank is a mathematical concept that is not necessarily in concordance with chemical information (Akbari Lakeh et al., 2017).

In what follows, the concordance of the local-rank information with the presence windows of components is pursued in a simulated bilinear chemical data set. The SMCR calculations for obtaining the Borgen plots and feasible solutions were made by the FAC-PACK software (version 1.20—the generalized Borgen plots module). This software can be freely downloaded at http://www.math.uni-rostock.de/facpack/.

Simulated example is a typical noise-free data, which is used for basically evaluating the concordance of the local-rank concept with the presence window of chemical species. The simulated data set is obtained from the chromatograms and spectra shown in Fig. 7.3, corresponding to a mixture of three co-eluting components in a chromatographic system monitoring with a diode array detector. The dimension of the simulated data set is 34 rows (elution times) by 51 columns (wavelengths). As it can be deduced from the elution and spectral profiles, there are three different local-rank windows on the rows of the data matrix comprising windows with rank one, two, and three.

Now, the concordance of the local-rank information with the presence windows of components in the simulated data is investigated. The agreement of the local-rank information with actual presence windows of the components in noise-free matrices is of great importance because the local-rank constraint is applied based on the correctness of this concordance. For instance, when selectivity is applied as a constraint, it is forced that just one component should participate to reconstruct the rows or columns with local-rank-one. The main question arises here: Is there any possibility that more than one component participates in building the local-rank-one window? Alternatively, in general, may rank deficiency be present in the subwindows of a data matrix?

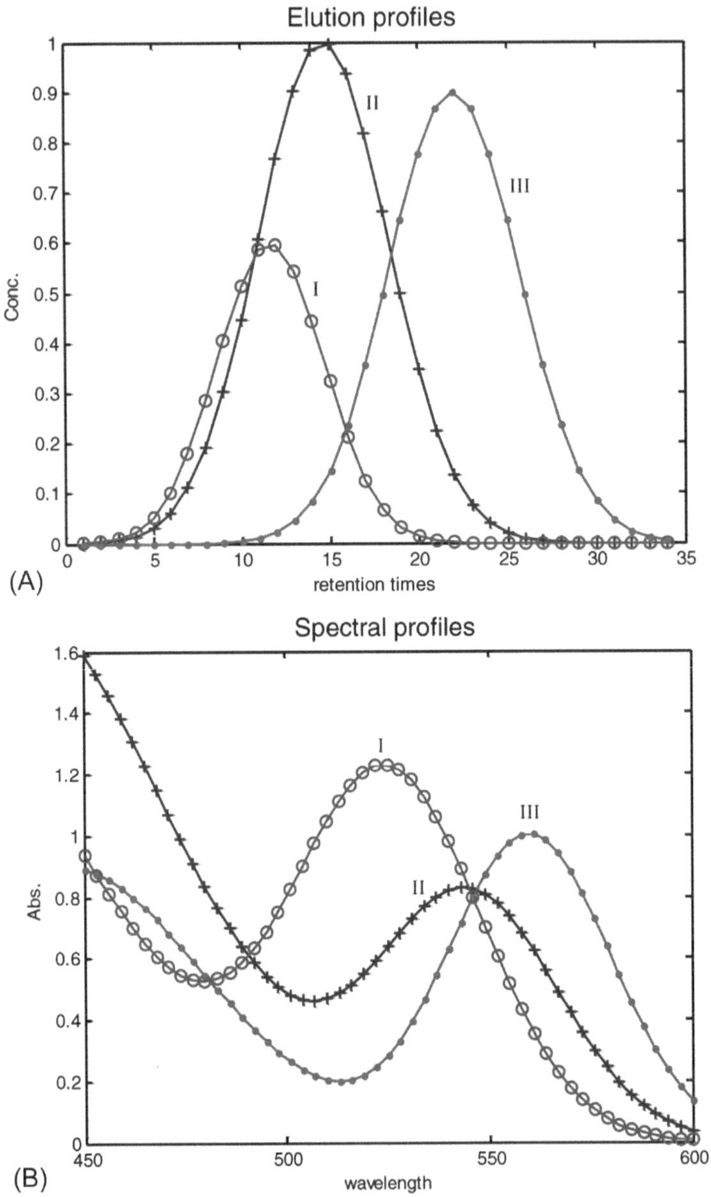

Fig. 7.3 (A) Simulated elution profiles; (B) simulated spectral profiles.

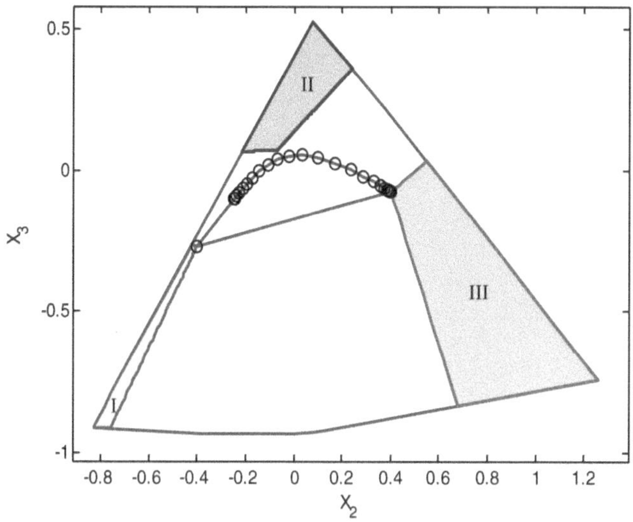

Fig. 7.4 Borgen plot of the simulated data set. The *hollow circles* indicate the rows of the data in row space.

Fig. 7.4 shows the Borgen plot of data in row space. In this plot, each circle is the representative of a spectrum at a specific retention time so the number of circles is equal to the number of rows of data. However, when some spectra have an identical shape, they will locate in the same position in row space. Three distinct feasible regions in Fig. 7.4 illustrate that rotational ambiguity exists for all three components. So, a range of feasible spectra can be defined for each component by non-negativity constraint. MCR methods usually try to find a set of feasible spectra that can construct data with corresponding non-negative concentration profiles. In the Borgen plot, a feasible set of solutions can be defined by any three points, each located in the feasible region of a component, that form a triangle that encloses all the data points (Henry, 2005; Rajkó, 2006). This will give a set of non-negative spectra and concentration profiles. Three typical sets of solutions are shown in Fig. 7.5A–C.

As seen in Fig. 7.6, different local-rank windows are present in the rows of the simulated data: the first eight spectra of the data set (correspond to the first eight elution times) make a window with rank-two, the next 14 spectra construct a window with rank three, spectra in rows 23–28 again make a window with rank two, and finally the last six rows build a window with rank one.

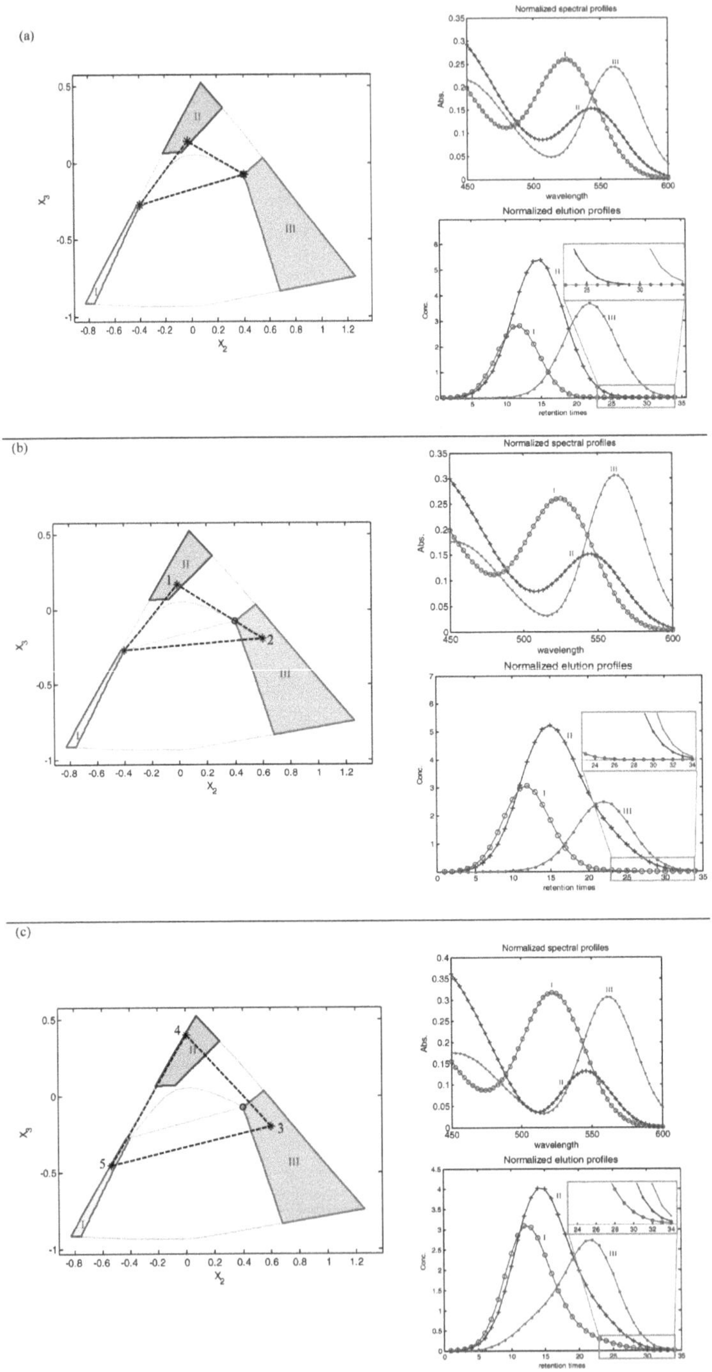

Fig. 7.5 The position of three possible solutions with different presence patterns. One of the possible set of solutions that (A) just one component, (B) two components, and (C) three components are present in the local-rank-one window.

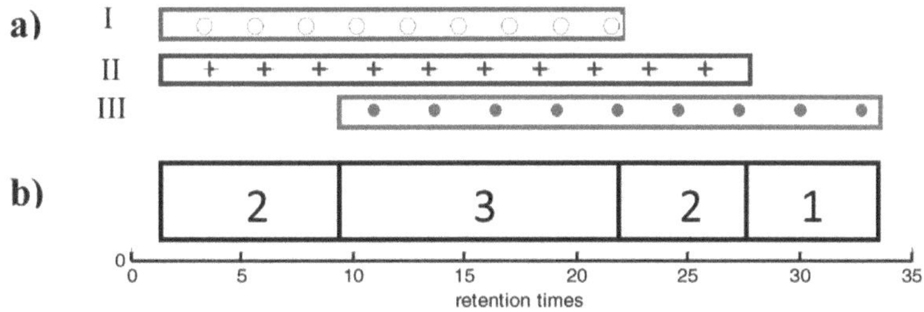

Fig. 7.6 (A) One of the possible elution chromatographic patterns among all for the simulated data set; (B) local-rank windows in rows of data matrix.

As stated before, the coordinates of spectra of the local-rank-one window would be the same in row space because all have the same shape. Thus, the six spectra of the local-rank-one window are indicated as a single circle in Fig. 7.5A–C; other data points are not shown for the sake of clarity. Regarding the Borgen plot in Fig. 7.5, the local-rank-one window is located inside the feasible region of the third component. Thus, the spectral profiles with rank one can be considered as the pure spectrum for the third component. Fig. 7.5A represents a set of feasible solutions that indicate the local-rank-one window as a selective window in data. However, one can find several sets of feasible solutions with this property by fixing the position of the third spectrum and changing the other two positions. As can be seen in elution profiles in Fig. 7.5A, there is a selective window for the third component (elution times 29–34 in arbitrary unit), and it shows that the contribution of other components for building the local-rank-one window is zero. However, other sets of feasible solutions may not confirm this conclusion. It is illustrated in Fig. 7.5B that a linear combination of the spectral profiles of the second and third component can construct the rows with rank-one, i.e., the linear combination of points 1 and 2 in Fig. 7.5B. The corresponding elution profiles confirm the presence of two components in the rank-one sub-matrix!

The local-rank-one window, in the same way, can be constructed from the linear combination of three different components, e.g., the points 3, 4, and 5 in Fig. 7.5C. Consequently, due to the rotational ambiguity, one, two, or three components may be present in this local-rank-one sub-matrix. Possibly previous chemical knowledge confirms the presence of more than one component in this window. In such cases, the number of recognizable contributions in a local window is lower than the number of chemical components, and for this reason that we call it "local rank deficiency." It is very important to note that all sets of feasible solutions of a system satisfy the local-rank information, but the presence patterns may differ from one set to the other because of local-rank deficiency. Thus, a number of various presence windows may exist in the feasible band of a component.

Therefore, if we have not any additional information just non-negativity, we have to use data-based uniqueness concept (Rajkó et al., 2015), and the forced improper application of Manne's theorems can cause artifact/biased results, sometimes to a critically high degree. That is why, the Manne's theorems are necessary but not sufficient for getting a unique resolution (Rajkó et al., 2015) unless with further information about the presence window of the components. If the presence window of each component is available, then information based on the zero concentration region can be used, and Manne's theorem can be successfully applied. For example, prior information about elution windows in case of chromatographic data or presence/absence of components in case of multiset data is valuable information to use in SMCR analysis in which applying Manne's theorem will produce a reliable result.

An efficient tool to study the conditions for achieving unique profile(s) by using zero concentration region information is principal component analysis (PCA) (Pearson, 1901; Tauler & Walczak, 2009). The possibility to express data information as basis sets of both row and column directions helps to understand the relationship between the abstract data spaces. On the other hand, there is a natural duality between the row and column vector spaces of the data matrix using minimal constraints. First time, Henry (2005) introduced the duality principle in chemometrics and described it for multivariate receptor modeling of compositional data of airborne pollution. Later Rajkó (2006) showed that there is a natural duality between the row and column vector spaces of a bilinear data matrix using only the non-negativity property of data set. He generalized the duality principle for the calculation of feasible solutions and also visualized it for a two-component system. It is remarkable that this mathematical relation between row and column spaces provides an efficient tool to transfer the information of the considered space to its dual space.

In our point of view, when there is the unique solution or conditions are suitable for obtaining the unique solution, the duality concept as a powerful approach can be used and illustrated in SMCR. In this section, it is assumed that reliable information about the presence and absence of components is available. It is shown how the adequate use of the zero concentration region information can produce the unique solution based on duality. In order to reach a better understanding of duality and its application in SMCR, mathematical formulae and geometrical visualization based on PCA are implemented.

A bilinear data can be decomposed by singular value decomposition (SVD) to the left eigenvectors (**U**), right eigenvectors (**V**), and a diagonal matrix of singular values (**S**). Therefore, if **U** and orthonormal **V** matrices define the basis vectors of the column and row spaces, respectively, according to duality, for any point in the considered space (e.g., V-space), there is a corresponding hyperplane in the dual space (U-space). Consequently, if the information is available to define a specific hyperplane, determination of the dual point is possible. Because of that, the conditions for achieving the unique solution can be studied simply based on this general principle.

Undoubtedly, the necessary condition to get the unique solution is the definition of the directed hyperplane in the dual space. Herein, based on the duality concept, we explore specifically how information about the presence of components can lead to a unique concentration or spectral profile for systems of any number of components. For obtaining the concentration profile of the analyte uniquely, the spectral subspace of the other components should be defined. This subspace is dual to the point in U-space, which results in the concentration profile of the analyte. An example is when there is a sub-matrix of M like M' whose rows has just information of all interfering compounds (the related presence window of all interfering compounds will not contain the analyte). By performing SVD, estimated loading vectors of M' can give the coefficients of the hyperplane in the dual space.

In any case, as it is depicted in Fig. 7.7, when the concentration feasible solutions of the green (gray in the printed version) and blue (light gray in the printed version) components are restricted to a plane, the red (dark gray in the printed version) one in V-space is uniquely resolvable. However, the reader should note that the zero concentration windows of the components should be known here. Moreover, it is worth to notice that the interferent hyperplane can be achieved when their pure profiles are given as well. Otherwise, one should estimate the zero concentration windows using local-rank information of the data matrix; however, the risk of this progress is discussed in an accompanying paper (Akbari Lakeh et al., 2017).

In the following section, zero-region information is implemented based on duality using a simulated HPLC-DAD data set. It gives an easy way for finding how to use zero concentration region information based on duality in special cases to obtain unique solutions using SMCR methods.

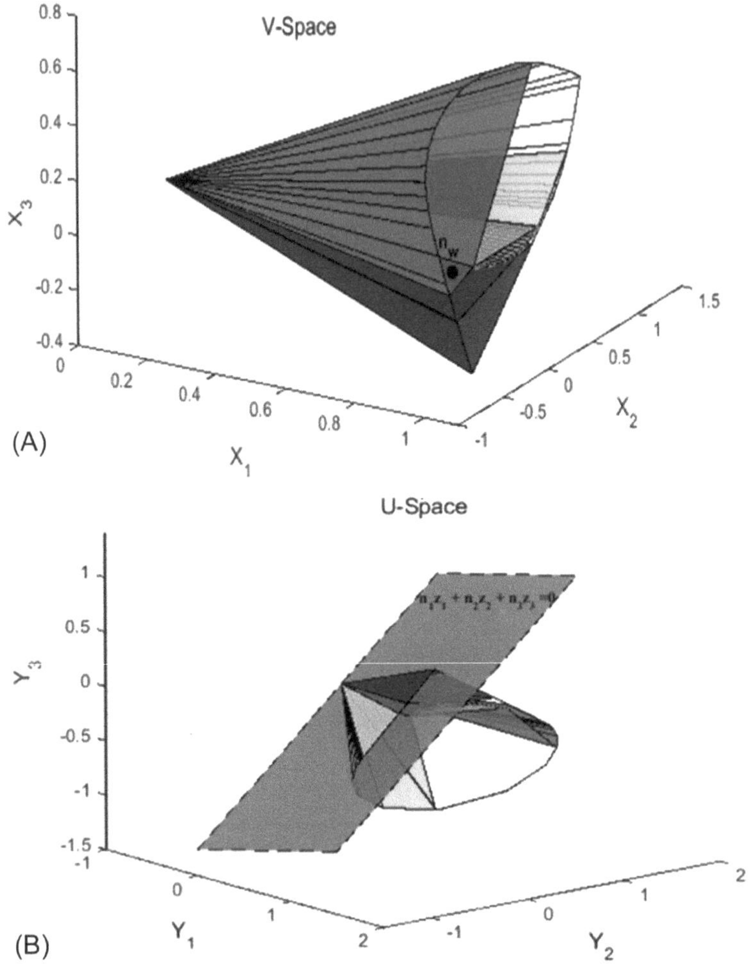

Fig. 7.7 Duality relation between the spectra of the *red (dark gray in the printed version)* component in V-space (A) the plane of *green (gray in the printed version)* and *blue (light gray in the printed version)* components in U-spaces (B).

Fig. 7.7 shows a typical HPLC-DAD data example. For clarifying the use of partial information in this section, the simulations and equations are presented for a three-component system. In this way, an appropriate HPLC-DAD data set is simulated by Gaussian functions. Fig. 7.8A–C displays the concentration profiles, spectral profiles, and the built-up data sets, respectively. The simulated data set is considered to be noise free to avoid complexity.

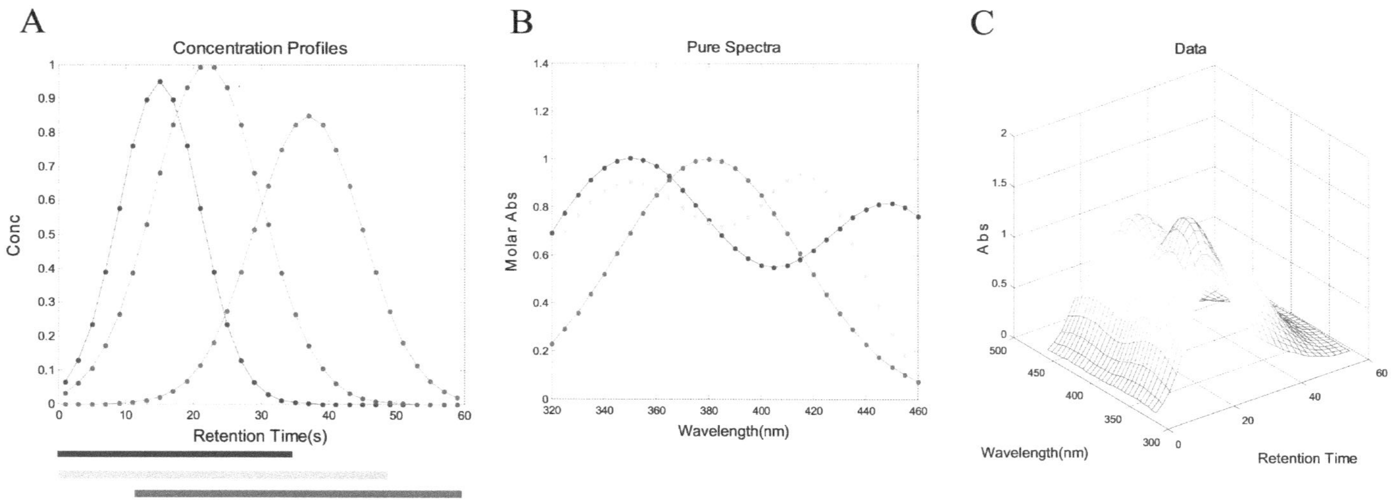

Fig. 7.8 Plots of (A) concentration profiles, (B) spectra profiles, (C) simulated data matrix.

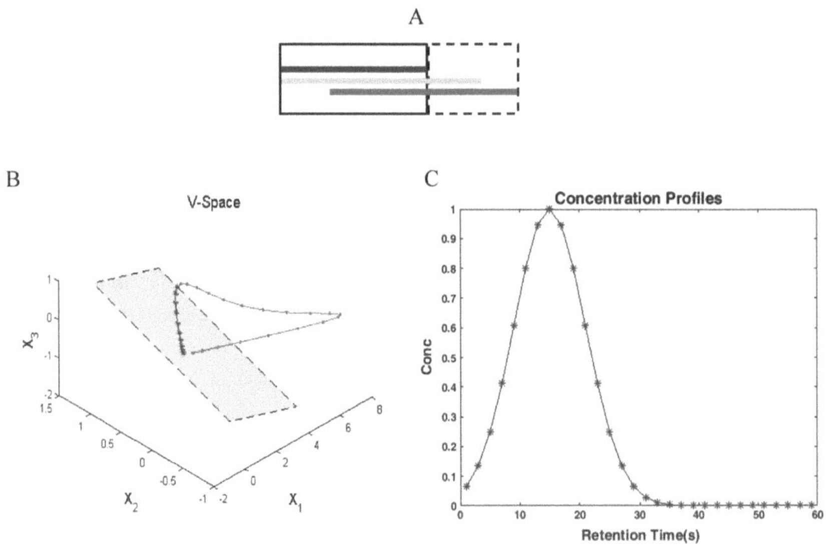

Fig. 7.9 (A) Depiction of concentration window of components. The *solid line rectangle* shows the concentration window of the *blue (light gray in the printed version)* component (analyte), and the *dashed line rectangle* shows its absence window. (B) Score plot of data matrix (*pink (gray in the printed version) dots*) and the spectral subspace of the *green (gray in the printed version)* and *red (dark gray in the printed version)* components (interferents shown with *blue (light gray in the printed version)* stars). (C) Calculated concentration profile for the *blue (light gray in the printed version)* component.

For obtaining the concentration profile of the analyte uniquely, the spectral subspace of the other components should be defined based on the duality relation. Herein, an example of such conditions is depicted for the simulated three-component system. Fig. 7.9A displays the concentration window of the components by horizontal lines along the time axis, and each line has the same color as the corresponding component in Fig. 7.8A. The blue (light gray in the printed version) component is considered as the analyte of interest, and two others are chosen as the interferents (the green (gray in the printed version) and red (dark gray in the printed version) ones). The solid line rectangle shows the concentration window of the blue (light gray in the printed version) component, and the dashed line rectangle shows its absence window. So in the simulated example, M′ is a sub-matrix of M associated with the dashed line rectangle, whose rows have just information of the green (gray in the printed version) and red (dark gray in the printed version) components. For calculating the requested plane, the loading vectors of M′ have to be calculated by performing SVD:

$$\mathbf{M}'_{i\times J} = \mathbf{U}'_{i\times 2}\,\mathbf{S}'_{2\times 2}\,\mathbf{V}'_{2\times J} \tag{7.18}$$

where i is the number of elution times of just the green (gray in the printed version) and red (dark gray in the printed version) components (excluding the blue (light gray in the printed version) one) and:

$$\mathbf{P}_{2\times 3} = \mathbf{V}'^{\mathbf{T}}_{2\times J}\,\mathbf{V}_{J\times 3} \tag{7.19}$$

The null space of matrix **P** is a vector, $\mathbf{n}_\mathrm{v}$, that will yield the coefficients of the demanded hyperplane:

$$n_1 z_{v1} + n_2 z_{v2} + n_3 z_{v3} = 0 \tag{7.20}$$

where z_v is a general variable vector, which represents the coordinates of an arbitrary point in V-space.

Based on Eq. (7.20), the spectral subspace of the green (gray in the printed version) and red (dark gray in the printed version) components is fixed in V-space (shown with cyan (gray in the printed version) color in Fig. 7.8B and the data points associated with these components lie on it). These points are shown with blue (light gray in the printed version) stars in the score plot of the data. The coefficients of the calculated plane correspond to a point in U-space. In fact, this point defines the coordinates related to the concentration profile of the blue (light gray in the printed version) component.

Since the elements of vector n_v are the coefficients of the plane equation, n_v can be transformed to the concentration profile of the blue (light gray in the printed version) component. Note that before transforming the coefficients of the plane, they have to be weighted by the corresponding singular values, i.e., the diagonal elements of the singular matrix of the data.

$$\mathbf{n}_{v\mathrm{w}(1\times 3)} = \mathbf{n}_{v(1\times 3)}\,\mathbf{S}_{(3\times 3)} \tag{7.21}$$

$$\mathbf{c}_{(\mathrm{blue})(1\times 1)} = \mathbf{U}_{(1\times 3)}\,\mathbf{n}^{\mathrm{T}}_{vw(3\times 1)} \tag{7.22}$$

The geometric visualization presented here would help to understand how the definition of the spectral subspace related to the interfering compounds leads to the unique concentration profile of the analyte. The correct detection of the subspace (with the described features), which can specify the spectral subspace of the interferents (green (gray in the printed version) and red (dark gray in the printed version)), leads to the unique analyte concentration profile (Fig. 7.9C). Also, the first Manne's theorem suggests these conditions for the unique resolution of the concentration profile. In the described example, these conditions are validated for the blue (light gray in the printed version) component, so it is possible to calculate its concentration profile uniquely. Therefore, it can be said that if the described necessary conditions are satisfied, then the spectral subspace of the interferents is determined, and by using the duality principle, the concentration profile of the analyte is uniquely resolvable. However, the reader should note here again that we have to know the zero concentration window of the blue (light gray in the printed version) component, and if it is unknown, estimating the zero concentration window using local-rank information of the data matrix would be risky (Akbari Lakeh et al., 2017).

Now it is clear which conditions using zero concentration region information of components lead to the unique solution in concentration and spectral spaces. As mentioned before, applying zero concentration information as selectivity or local-rank constraints can reduce the rotational ambiguity, but they do not necessarily result in the unique solution. For providing a unique solution, the available zero region information has to define hyperplane of all interfering compounds in the dual space. In the cases when unique solution is not provided, information is not enough to construct the hyperplane of interfering components.

Again, we should mention here as well that we should have prior information about the green (gray in the printed version) component: in both its concentration windows, only one interferent can be present and the other should be absent, respectively. Otherwise, we should estimate these concentration windows using local-rank information of the data matrix; however as was mentioned above, the risk of this treatment can be emerged.

5. Equality constraint

Equality is one of the most important constraints to restrict the feasible solutions and reach to unique solutions in SMCR methods. Imposing equality constraint, in the form of information about a known specific profile, results in a fixed subspace for the complementary components in the dual space. The implementation of the knowledge of a known profile as an equality constraint and its significant effects on the rotational ambiguity has been investigated by considering theoretical and practical aspects of the constraint (Beyramysoltan, Abdollahi, & Rajkó, 2014; Ghaffari & Abdollahi, 2018; Sawall, Fischer, Heller, & Neymeyr, 2012). Equality is a powerful constraint to restrict the feasible solutions and reach to unique solutions in SMCR methods. Analytical implementation of equality constraint in two-component data is very simple due to fixing the coordinates of known profile and complementary dual profile in their corresponding abstract subspaces.

The idea of the analytical approach to implement equality constraint for Borgen plot is inspired by the geometry of the abstract space for data sets of three-component systems that have one unique profile. The procedure to determine the new abstract geometry after imposing the equality constraint has four steps. The schematic of the algorithm is shown in Fig. 7.10.

Where the spectral profile of the third component as an example of a three-component data is known. For simplicity, the spectral space in the schematic was supposed as the space containing the coordinate of the unique profile; however, of course, the space can be the concentration space where a concentration profile is known. The analytical methods are fast enough to calculate and draw the abstract spaces and feasible bands in less than 1 s using an average commercial computer. The analytical SMCR methods are limited to use on those systems that have moderate noise only, and they are very useful to investigate the microscopic structures of chemical data and visualize the effects of several constraints and limitations on the accuracy of bilinear curve resolution methods.

The effect of equality constraints on the reduction of the rotational ambiguity can be very different. The relative position of the known component in the area of feasible solution (AFS) explains the behavior. A measure is suggested that indicates the impact of the equality constraints (Sawall et al., 2020). The representation of this measure in the area of feasible solutions shows strong variations in the restrictiveness of equality constraints. This characteristic number can also be considered as restrictiveness measure. A relatively large reduced AFS shows that the equality constraint cannot significantly reduce the rotational ambiguity. The profile that is locked by the equality constraint is compatible to many other solutions. In contrast to this, a small reduced AFS means that the choice of the equality constraint is very restrictive for the factorization problem. An equality constraint strongly reduces the AFS if it has a high restrictiveness.

Similarly, a weak AFS reduction corresponds to a low restrictiveness. A restrictiveness estimator is based on the volume of the reduced AFS under the equality constraint. The volume of the reduced AFS under the equality constraint is a measure of restrictiveness named $\kappa 1$. The ambiguity estimators not only calculate for single points, but compute the estimators for every point in the AFS. The numerical estimator values are visualized along a color

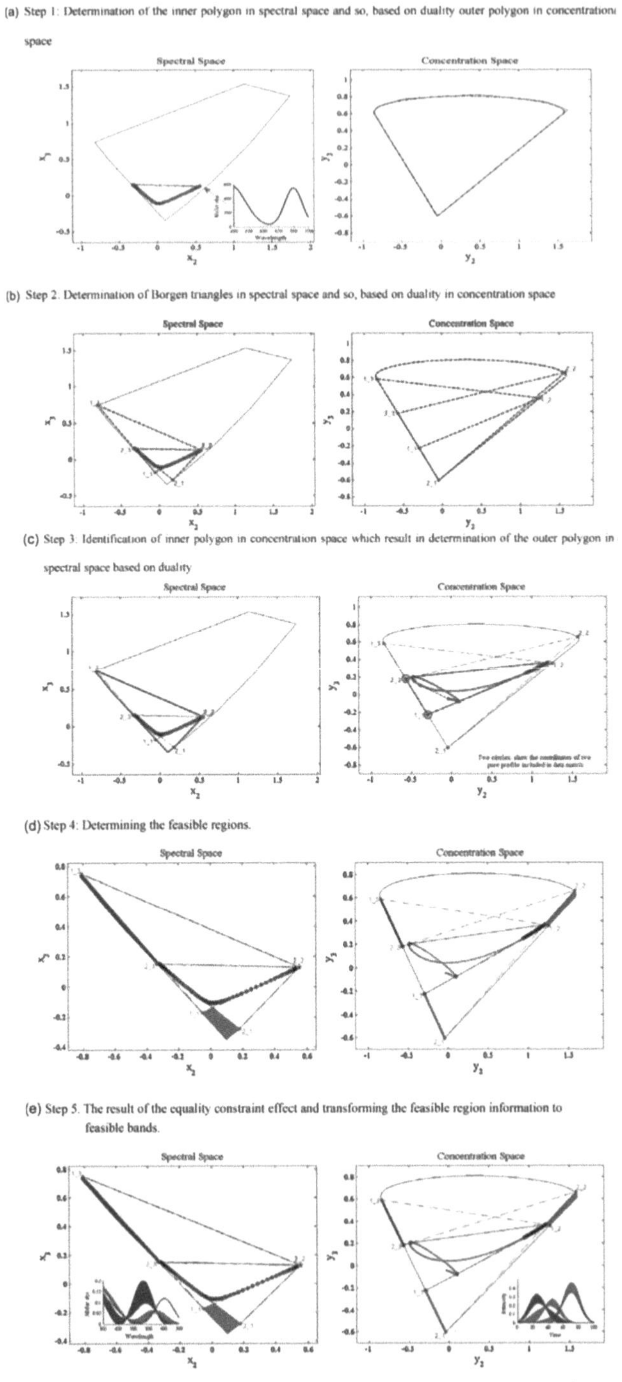

Fig. 7.10 Demonstration of the steps of how to impose the equality constraint in Borgen plot (A–D); (E) represents the new arrangement obtained by imposing the equality constraint.

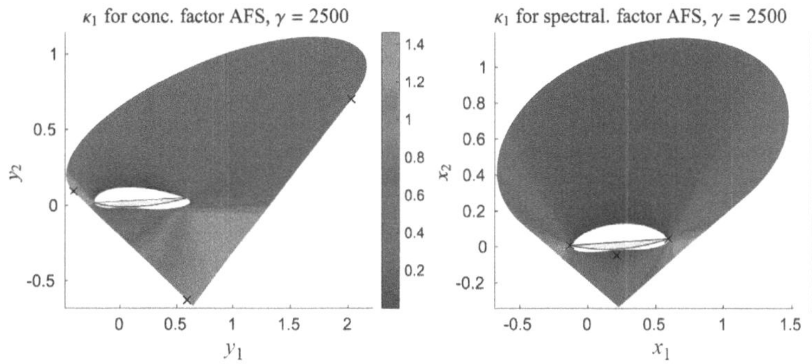

Fig. 7.11 Behavior of the AFS-volume-based ambiguity estimator κ1 for the model problem.

coordinate. In other words, the full area of the two-dimensional AFS is colored by the values of κ1. A color legend for each plot enables a size assessment of the estimator values. The resulting AFS plots have a significantly more in-depth information content compared to uniform color AFS plots. However, the computational effort is large. Fig. 7.11 shows the behavior of the AFS-volume-based ambiguity estimator κ1 for the concentration factor AFS and for the spectral factor AFS.

Rotational ambiguity in SMCR methods can be affected by some other profile shape-based constraints such as unimodality and monotonicity. The monotonicity and unimodality as meaningful, chemistry-based constraints are used for the reduction of the rotational ambiguity underlying multivariate curve resolution methods. Both of these constraints can also produce unique profiles in chromatographic, kinetics, and chemical equilibria systems (Zade et al., 2019; Zade, Neymeyr, Abdollahi, & Sawall, 2021).

6. Hard models constraint

In evolving systems where concentration changes over time according to chemical kinetics or chemical equilibria, physical-chemical constraints (the so-called hard modeling) can be used in a powerful way to calculate concentration profiles by imposing the mathematical functions of the chemical model as constraints. These kinds of constraints generally reduce the range of feasible solutions and often result in unique solutions for components that can be described with a proper physical-chemical model, provided the initial conditions are precisely known (Beyramysoltan et al., 2014). In this section, the effects of a hard modeling constraint on the reduction of the extent of rotational ambiguity have been investigated (Golshan, Abdollahi, & Maeder, 2012; Khodadadi Karimvand & Abdollahi, 2018).

A simulated data set was analyzed. The kinetic model used for simulated data was the irreversible parallel reactions in the form of $A \xrightarrow{K_1} B; F \xrightarrow{K_2} D$, assuming that initially only species A and F are present in the reaction mixture. The four kinetic equations, Eqs. (7.23)–(7.26), can be used to describe the concentration profiles of the reactants (A, F) and the products (B, D) of the considered two first-order reactions.

$$[A]_i = [A]_0 \exp(-k_1 t_i) \quad (7.23)$$

$$[B]_i = [A]_0 (1 - \exp(-k_1 t_i)) \quad (7.24)$$

$$[F]_i = [F]_0 \exp(-k_2 t_i) \quad (7.25)$$

$$[D]_i = [F]_0 (1 - \exp(-k_2 t_i)) \quad (7.26)$$

Also to remain in the realm of three-component systems, we assume that F has no absorption in the spectral region investigated. Concentration profiles for the reactants and the product were simulated based on chemical models (Eqs. 7.23–7.26), using the rate constants equal to $k_1 = 0.05\,\text{s}^{-1}$ and $k_2 = 0.15\,\text{s}^{-1}$, with $[A]_0 = [F]_0 = 1\,\text{M}$. Measurement times were between 1 and 47 or 4 and 47 s. The simulated absorbance spectra of data set **X** were then calculated by post-multiplication of concentration profiles with pure spectra. To more rigorously test the methods, random errors by normal distributions with mean zero and standard deviation equal to 0.3% of averaged absorbance were added to the simulated data. The concentration and spectra profiles and the resulting "measured" spectra for data set **X** are plotted in Fig. 7.12. It contains 24 spectra corresponding to the calculated concentration profiles. The measurement time for this data has been chosen between 1 and 47 with an increment of 2. If spectra starting at time 0 are included, the rotational ambiguity is very small.

The grid search method performs the calculation of all feasible solutions fitting the measured data and fulfilling the constraints of non-negativity and the hard model, also delivering the rate constants for the modeled reaction.

Fig. 7.10A shows the areas of feasible solutions which were obtained independently by the grid search minimization of the residuals procedure in U-space under non-negativity (in brown (gray in the printed version)) and after additional hard constraints of the same data set **X** in red (dark gray in the printed version) for D and blue (light gray in the printed version) and green (gray in the printed version) for A and B.

The introduction of the hard constraint results in perfectly defined concentration profiles for A and B and reduced AFS for the concentration

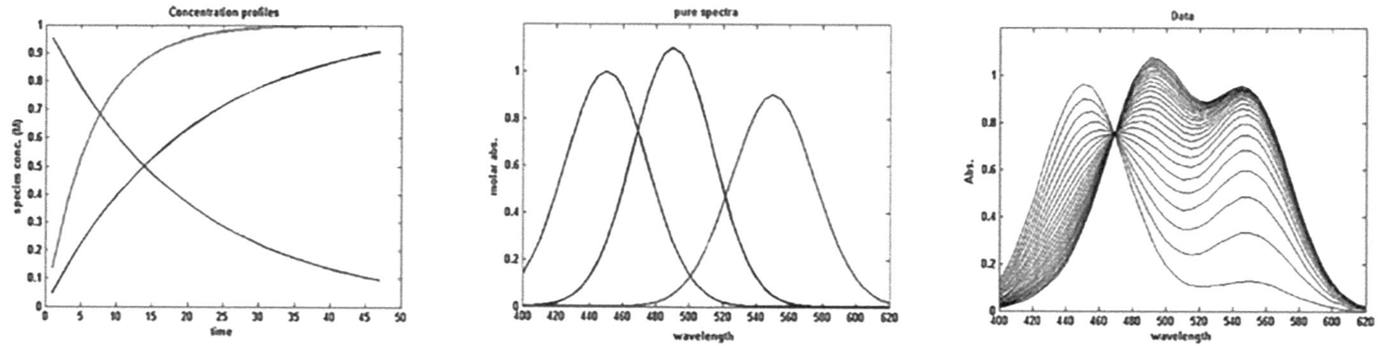

Fig. 7.12 Concentration profiles **C**, absorption spectra **S**, and data matrix **X** for the simulation.

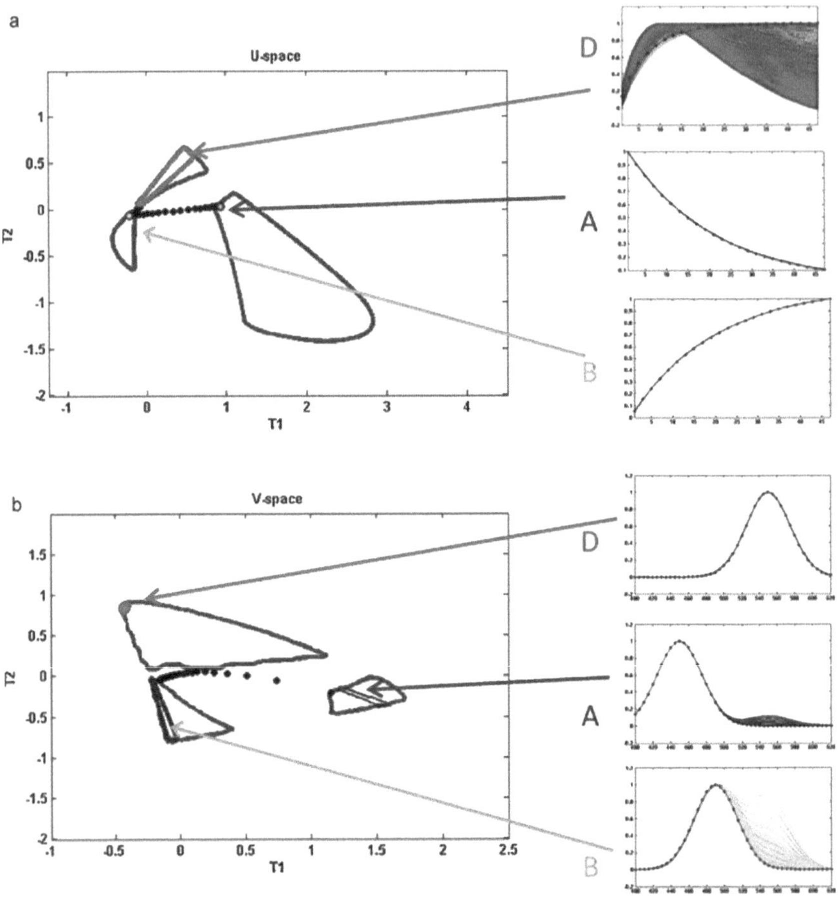

Fig. 7.13 (A) Area of feasible solutions (AFS) for simulated data set **X** in U-space by using non-negativity and hard constraint in the presence of noise. (B) AFS for data set **X** in V-space by using non-negativity and hard constraint in the presence of noise. The *brown (gray in the printed version) contour plots* are calculated by using only non-negativity constraint, and the *black circles* are the distribution of the data matrix in both spaces. (A) Range of feasible concentration profiles. (B) Range of feasible absorption spectra. Each concentration profile and absorption spectrum is normalized to a maximum of one.

profile of *D*. Fig. 7.13B displays the equivalent V-space. At first sight, it might be surprising that the spectrum of the soft modeled component *D* is uniquely defined, while the AFS for the components *A* and *B* are only reduced but not unique.

The results of the simulation study show that the additional hard model successfully resolves the feasible ranges of the components that take part in chemical model. In addition, the feasible region of one unknown component was reduced. By incorporating a hard modeling constrain to soft modeling methods, and incorporating knowledge about some of the species, generally superior results can be obtained.

It should be noted that the hard modeling constraint acts like the equality constraint and the rules of both are comparable. Only, exact mathematical functions are placed instead of a specific profile.

6.1 Spectral curve-fitting constraint

Hard modeling constraints can also be applied for spectral profiles if mathematical functional forms are imposed (peak-shaped models). Spectral hard modeling is also known as curve fitting or peak fitting. Curve-fitting methods have an old history in the literature and are used for qualitative or quantitative resolution of systems. In these methods, spectra are fitted to sums of peak-shaped curves, e.g., Gaussian or Lorentzian shapes or combinations of Gaussian and Lorentzian functions, which are known as Voigt functions (Thorne, 1974).

The effect of spectral hard modeling constraints on the feasible solutions of soft methods for simulated and real data sets was investigated. For each data set, the area of feasible solutions (AFS) were calculated. The AFS were transformed into real profiles and fitted to the summation of Voigt functions. All points in the feasible region fit the model well; thus, one can conclude that the peak-shaped models did not reduce the range of feasible bands. As further evidence of the nonuniqueness results of spectral fitting, the data sets were adequately fitted to different Voigt peak models with the minimum number of peaks that gave a statistically sufficient model.

Achieving an accurate, precise model of concentration profiles with hard modeling methods requires a significant amount of chemical information about the studied system. When sufficient chemical information is known, applying models for concentration profiles in SMCR methods can result in real, unique solutions. Lacking this type of information in the peak-shaped models of spectra results in non-unique solutions for spectral fitting. There are laws that govern the shapes of concentration profile models, i.e., the law of mass action, but there are no strong laws about spectra. We can only approximate them with a selection of Gaussian, Lorentzian, or Voigt peak functions. Applying a Voigt peak-shaped model without additional information as constraints on spectroscopic signals does not have any effect on rotational ambiguity in resolving bilinear data.

7. General rule for uniqueness (GRU) (Karimvand et al., 2020)

In a research, the duality concept is applied to propose a general rule for unifying all the different information and constraints that can produce a unique solution in SMCR methods. This general rule for uniqueness (GRU) describes how different information about a chemical component can help to resolve its corresponding profile uniquely. Moreover, GRU provides a better understanding about the uniqueness concept and can be a guide for developing new constraints in SMCR to get more accurate solutions.

The duality concept is a powerful tool in explaining and interpreting numerous features of MCR methods. Based on this concept, a general rule was employed to explain the unique solutions of MCR methods as a result of applying different constraints. The GRU states that any information able to fix the subspace of all complementary components in one space results in a unique solution for the analyte in the other space. This generalization is presented for constraints such as trilinearity, equality, zero concentration region, correspondence, local rank, non-negativity under DBU conditions, and even to extract the NAS. GRU is a proper index to find the conditions and information in different chemical systems probably leading to a unique solution in applying MCR methods. Hence, by fulfilling the proposed general rule of uniqueness, it is possible to successfully introduce new constraints and apply different conditions to achieve a unique solution.

7.1 Generalized rule for uniqueness (GRU)

Based on the Bouguer-Lambert-Beer law, a spectroscopic data set can be written as a product of two matrices, C and A, containing the concentration and spectral profiles of individual species, respectively. Hence, a bilinear decomposition of a given data set is:

$$\mathbf{D} = \mathbf{C}\,\mathbf{A}^{\mathrm{T}} \tag{7.27}$$

where **D** is the measured data matrix with I rows (observations) and J columns (variables). Additionally, a bilinear data can be decomposed by singular value decomposition (SVD):

$$\begin{aligned} \mathbf{D} &= \mathbf{U}\mathbf{S}\mathbf{V}^{\mathrm{T}} = \mathbf{X}\mathbf{V}^{\mathrm{T}} = \mathbf{U}\mathbf{Y}^{\mathrm{T}} \geq \mathbf{0} \\ \mathbf{D}^{\mathrm{T}} &= \mathbf{V}\mathbf{S}\mathbf{U}^{\mathrm{T}} = \mathbf{Y}\mathbf{U}^{\mathrm{T}} = \mathbf{V}\mathbf{X}^{\mathrm{T}} \geq \mathbf{0} \end{aligned} \tag{7.28}$$

In Eq. (7.28), **U** and **V** represent the left and right eigenvectors of $\mathbf{D}_{(\mathrm{I}\times\mathrm{J})}$, respectively, and **S** is a diagonal matrix of singular values. Also, **X** and **Y** are the coordinates of the row and column vectors of the data set in the row (V-space) and column space (U-space); in turn, 0 is a matrix containing zeroes with the suitable dimension ($I \times J$ or $J \times I$). Moreover, Eq. (7.28) shows a natural duality between row and column vectors of a data matrix using the only non-negativity property of the data set. Each point defined by t in the considered space conforms to the duality relation. Thus, it is easy to derive the coefficients to create the hyperplane corresponding to this point in another space (dual space).

$$\mathbf{n} = \mathbf{t}\,\mathbf{S}^{-1} \tag{7.29}$$

where **t** represents the coordinates of a fixed point, **S** is the matrix of singular values, and **n** is the hyperplane coefficients vector. Point **t** in one space of matrix has a certain hyperplane defined by Eq. (7.30) in another space (dual space).

$$\mathbf{n}^{\mathrm{T}}\,\mathbf{z} = 0 \tag{7.30}$$

In the equation of hyperplane, **z** is a general variable vector representing the coordinates of an arbitrary point either in V- or in U-spaces and 0 is a scalar.

Hence, a certain point in the abstract space corresponds to a unique hyperplane in its dual subspace. A unique profile solution in the result of an MCR method is equivalent to a certain point in the corresponding abstract vector space of the profile. In fact, the duality concept states that a proper hyperplane should be defined in the dual space to achieve a unique solution for an analyte (the correct and sufficient meaningful physical/chemical information can define the proper and correct hyperplane corresponding to true solution). Therefore, accurate information must be used to determine the hyperplane as the subspace related to all complementary compounds (this specific hyperplane is a subspace of all complementary (interfering) compounds without the contribution of the analyte). Thus, the information of the proper constraints resulting in a unique solution in the vector space of the data should lead to fix a hyperplane in its dual space. Therefore, a general rule for the uniqueness of MCR solutions can be proposed based on the duality concept:

The necessary condition to get a unique solution for the analyte is the possibility of the definition of the hyperplane related to all interfering components in the dual space by implementing adequate physical/chemical constraints (or information).

Indeed, for a (N) component system, a unique solution always fixes a point in one space and an (N-1)-dimensional hyperplane of complement components in the dual space. This is a general rule of uniqueness in SMCR methods.

Herein, with the help of a geometrical approach, and for a generated three-component system, it is observed that all the above-mentioned constraints (or information) can provide a unique solution by correctly determining the hyperplane of all interfering compounds.

References

Abdollahi, H., & Nazari, F. (2003). Rank annihilation factor analysis for spectrophotometric study of complex formation equilibria. *Analytica Chimica Acta*, *486*(1), 109–123.

Abdollahi, H., & Sajjadi, S. (2010). On rotational ambiguity in parallel factor analysis. *Chemometrics and Intelligent Laboratory Systems*, *103*(2), 144–151.

Akbari Lakeh, M., Rajkó, R., & Abdollahi, H. (2017). Local rank deficiency caused problems in analyzing chemical data. *Analytical Chemistry*, *89*(4), 2259–2266.

Alier, M., Felipe, M., Hernández, I., & Tauler, R. (2011). Trilinearity and component interaction constraints in the multivariate curve resolution investigation of NO and O 3 pollution in Barcelona. *Analytical and Bioanalytical Chemistry*, *399*(6), 2015–2029.

Beyramysoltan, S., Abdollahi, H., & Rajkó, R. (2014). Newer developments on self-modeling curve resolution implementing equality and unimodality constraints. *Analytica Chimica Acta*, *827*, 1–14.

Beyramysoltan, S., Rajkó, R., & Abdollahi, H. (2013). Investigation of the equality constraint effect on the reduction of the rotational ambiguity in three-component system using a novel grid search method. *Analytica Chimica Acta*, *791*, 25–35.

Cattell, R. B. (1944). "Parallel proportional profiles" and other principles for determining the choice of factors by rotation. *Psychometrika*, *9*(4), 267–283.

Ghaffari, M., & Abdollahi, H. (2018). Duality based interpretation of uniqueness in the trilinear decompositions. *Chemometrics and Intelligent Laboratory Systems*, *177*, 17–25.

Golshan, A., Abdollahi, H., & Maeder, M. (2012). The reduction of rotational ambiguity in soft-modeling by introducing hard models. *Analytica Chimica Acta*, *709*, 32–40.

Harshman, R. A. (1970). Foundations of the PARAFAC procedure: Models and conditions for an" explanatory" multimodal factor analysis. *UCLA working papers in phonetics, 16* (pp. 1–84). Ann Arbor, Michigan: University Microfilms. No. 10,085.

Henry, R. C. (2005). Duality in multivariate receptor models. *Chemometrics and Intelligent Laboratory Systems*, 77(1–2), 59–63.

Karimvand, S. K., Lakeh, M. A., Tavakkoli, E., Ghaffari, M., Omidikia, N., Abad, S. K. A., et al. (2020). A general rule for uniqueness in self-modeling curve resolution methods. *Journal of Chemometrics*, *34*(7), e3268.

Khodadadi Karimvand, S., & Abdollahi, H. (2018). Chemometrical study of spectral curve fitting constraint on self-modelling curve resolution methods. *Journal of Chemometrics*, *32*(12), e3074.

Lawton, W. H., & Sylvestre, E. A. (1971). Self modeling curve resolution. *Technometrics*, *13*(3), 617–633.

Manne, R. (1995). On the resolution problem in hyphenated chromatography. *Chemometrics and Intelligent Laboratory Systems*, *27*(1), 89–94.

Pearson, K. (1901). LIII. On lines and planes of closest fit to systems of points in space. *The London, Edinburgh, and Dublin Philosophical Magazine and Journal of Science*, *2*(11), 559–572.

Rajkó, R. (2006). Natural duality in minimal constrained self modeling curve resolution. *Journal of Chemometrics: A Journal of the Chemometrics Society*, *20*(3–4), 164–169.

Rajkó, R. (2010). Comments on "Near-infrared hyperspectral unmixing based on a minimum volume criterion for fast and accurate chemometric characterization of counterfeit tablets". *Analytical Chemistry*, *82*(20), 8750–8752.
Rajkó, R., Abdollahi, H., Beyramysoltan, S., & Omidikia, N. (2015). Definition and detection of data-based uniqueness in evaluating bilinear (two-way) chemical measurements. *Analytica Chimica Acta*, *855*, 21–33.
Sanchez, E., & Kowalski, B. R. (1986). Generalized rank annihilation factor analysis. *Analytical Chemistry*, *58*(2), 496–499.
Sawall, M., Fischer, C., Heller, D., & Neymeyr, K. (2012). Reduction of the rotational ambiguity of curve resolution techniques under partial knowledge of the factors. Complementarity and coupling theorems. *Journal of Chemometrics*, *26*(10), 526–537.
Sawall, M., Zade, S. V., Kubis, C., Schröder, H., Meinhardt, D., Brächer, A., et al. (2020). On the restrictiveness of equality constraints in multivariate curve resolution. *Chemometrics and Intelligent Laboratory Systems*, *199*, 103942.
Sidiropoulos, N. D., & Bro, R. (2000). On the uniqueness of multilinear decomposition of N-way arrays. *Journal of Chemometrics*, *14*(3), 229–239.
Stegeman, A., & Sidiropoulos, N. D. (2007). On Kruskal's uniqueness condition for the Candecomp/Parafac decomposition. *Linear Algebra and its Applications*, *420*(2–3), 540–552.
Tauler, R., & Walczak, B. (2009). *Comprehensive chemometrics: Chemical and biochemical data analysis*. Elsevier.
Tavakkoli, E., Rajkó, R., & Abdollahi, H. (2018). Duality based direct resolution of unique profiles using zero concentration region information. *Talanta*, *184*, 557–564.
Thorne, A. P. (1974). *Spectrophysics*. London: Chapman and Hall.
Zade, S. V., Neymeyr, K., Abdollahi, H., & Sawall, M. (2021). Characterization of the unimodality constraint as an effective chemistry-based condition in resolving of chemical processes data. *Microchemical Journal*, *160*, 105615.
Zade, S. V., Sawall, M., Neymeyr, K., & Abdollahi, H. (2019). Introducing the monotonicity constraint as an effective chemistry-based condition in self-modeling curve resolution. *Chemometrics and Intelligent Laboratory Systems*, *190*, 22–32.

APPENDIX

Introduction to python

Parvin Razzaghi

1. Introduction

Python is a popular programming language, which is created by Guido van Rossum (Van Rossum & Drake Jr, 1995a, 1995b) and released for the first time in 1991 as Python 0.9.0. Python for the first time is created as a successor to the ABC programming language in the late 1980s. Python is an interpreted object-oriented high-level general-purpose programming language. From 1991 until now, 27 versions are released. The latest version of python is python 3.10, which is released on October 4, 2021 (Salgado, 2021).

Python is popular, and in many cases, it is the first choice of the programming language for users. This popularity has many reasons, and some of these reasons are mentioned in the following:

1. Python is easy to learn because its syntax is simple and commands are designed such that it is an English-like language. Also, python is so flexible and lets developers develop any sort of applications that they want.
2. Python has an active support community. It helps users (beginners to experts) to access many sources of documentation and tutorials. This property helps the community to grow much better. Also, python uses as a popular programming language in the academic community.
3. One of the other reasons is that python provides many libraries and frameworks for many research areas. Hence, it is suitable for researchers to form a diverse community. Also, python provides many useful libraries for data science machine learning and cloud computing. Since these research areas get more attention in recent years, these python libraries receive more attention from users.
4. Python programming language is supported by big companies like Amazon web services, Facebook, and Google. It leads the python package to grow faster.

Up to now, many useful packages are developed for many research areas. In chemoinformatics and bioinformatics too, many useful packages like ChemoPy, PyDPL, DeepChem, sklearn, and pyMCR are developed. In this chapter, these packages are introduced and discussed. Then, the python codes of multivariate algorithms, which are investigated in Chapter 3, are given.

2. ChemoPy

ChemoPy is a freely available python package, which is introduced by Cao, Xu, Hu, and Liang (2013). Many applications in chemoinformatics need to provide a discriminative representation for a small molecule. Also, there are many molecular representation algorithms in the literature: ChemoPy provides seven types of molecular fingerprint systems for drug molecules. These types include topological fingerprints, electro-topological state (E-state) fingerprints, MACCS keys, FP4 keys, atom pairs fingerprints, topological torsion fingerprints, and Morgan/circular fingerprints. Also, this package could provide a large number of 3D molecular descriptors by applying a semi-empirical quantum chemistry program MOPAC (Stewart, 1990). The last version of this package is released on January 31, 2013, and it runs on MS Windows and Linux. This package could be accessible from this link https://code.google.com/archive/p/pychem/downloads.

3. PyDPI

PyDPI (drug-protein interaction with Python) is another free available python toolkit that is provided for drug discovery and is useful in chemoinformatics, bioinformatics, chemogenomics, and proteochemometrics (Cao, Liang, et al., 2013). This package is available via https://sourceforge.net/projects/pydpicao/.

PyDPI provides the useful known state-of-the-art descriptors for protein sequence, drug molecules, protein-protein interaction network, and protein-ligand interaction network. PyDPI computes six feature groups for protein sequence including amino acid composition (consists of three features amino acid composition, dipeptide composition, and tripeptide composition features), autocorrelation (consists of normalized Moreau-Broto autocorrelation, Moran autocorrelation, and Geary autocorrelation features), CTD (consists of composition, transition, and distribution features), conjoint triad, quasi-sequence order sequence (consists of order coupling number and quasi-sequence order descriptors), and pseudo-amino acid composition (consists of pseudo-amino acid composition and pseudo-amino acid composition features). Also, it computes 13 feature groups for small molecules, including the constitution, topology, connectivity, *E*-state, kappa, burden, information, autocorrelation, charge, property, MOE type, and fingerprints. Also, there are, respectively, three and two methods to construct the interaction descriptor for protein-protein and protein-ligand interaction networks.

4. pyMCR

Camp Jr (2019) has introduced a python library for multivariate curve resolution analysis with alternating regression, which is called pyMCR. This software library, written in Python, enables users to perform multivariate curve resolution (MCR) analysis, which lets users choose the error functions for minimization, constraints, and regressors.

5. DeepChem

DeepChem is an open-source framework, which is first created by Bharath Ramsundar at Stanford University. Now, it is developed by a group of developers; however, anyone could join and contribute. This project uses TensorFlow to create deep learning models. It is developed to facilitate the researchers to develop new deep learning-based algorithms in drug discovery, biology, and life science. DeepChem provides many useful APIs to develop algorithms. These APIs include (1) data (for handling data) and (2) features: extract features from input data to feed into machine learning models. DeepChem provides both data-driven-based approaches like graph convolution feature and non-data-driven-based approaches like CircularFingerprint, PubChemFingerprint, Mol2VecFingerprint, and RDKitDescriptors. (3) Scikit-learn models, (4) Keras models, (5) PyTorch models, (6) Transformers: provide transformer object to adopt the mechanism of attention by weighting each part of the input data to emphasize the important parts, (7) hyperparameter tuning, and (8) reinforcement learning: reinforcement learning is a powerful technique in machine learning: DeepChem provides a useful framework to use it to solve computational biology problems. Also, there are many other beneficial APIs and the suitable documentation about them is available in https://deepchem.readthedocs.io/en/latest/.

6. Scikit-learn

Scikit-learn is a machine learning package in python, which is open source and commercially usable under the BSD license. This package contains many useful objects, including preprocessing, classification, regression, dimensionality reduction, clustering, and model selection. The last version of this package is scikit-learn 1.0.1, which is released in October 2021. It is built on NumPy, SciPy, and matplotlib and provides many useful tools for visualizing data.

7. Examples

In this section, three examples (dimensionality reduction, partial least-squares (PLS), and clustering analysis) are provided using "sklearn" package.

7.1 Dimensionality reduction

In this section, dimensionality reduction using principal component analysis (PCA) is explained. In dimensionality reduction, the main goal is to project the input data into a new space with low dimensions such that it preserves the most important knowledge of the input data. Dimensionality reduction approaches are divided into two categories: feature selection and feature projection (i.e., feature extraction). In both categories, there are many algorithms. In this section, PCA from the feature projection category is practically explained. To this end, a toy dataset is selected, and then, the steps of how PCA is applied to it to reduce the dimension are explained.

In python, at the first step, you should import the necessary libraries to run the code. In this example, we need "numpy," "sklearn" (to use PCA and load one of the machine learning toy datasets), and "matplotlib" (to show the results) libraries.

```
import numpy as np
from sklearn import datasets
from sklearn.decomposition import PCA
import matplotlib.pyplot as plt
```

In the next step, choose the toy dataset that you want to apply the PCA on. In this section, we select the IRIS dataset that contains three classes of 150 samples with four features.

```
from sklearn.datasets import load_iris
data = load_iris()
```

Next, a PCA object is created and it is fit on data by the following commands:

```
from sklearn.decomposition import PCA
pca = PCA(n_components=2)
data_reduced = pca.fit_transform(data.data)
```

It should be noted that PCA class gets an argument named "*n_components*," which determines the dimension of the new space. Now, to evaluate how it is performed on data, the projected data is visualized by "matplotlib" package.

```
import matplotlib.pyplot as plt
idx = np.argwhere(data.target==0)
plt.scatter(data_reduced[idx,0],data_reduced[idx,1],color='blue')
idx = np.argwhere(data.target==1)
plt.scatter(data_reduced[idx,0],data_reduced[idx,1],color='red')
idx = np.argwhere(data.target==2)
plt.scatter(data_reduced[idx,0],data_reduced[idx,1],
 color='purple')
plt.gca().set(title="PCA on IRIS Dataset",
xlabel='Feature1', ylabel='Feature2')
plt.legend()
plt.show()
```

By running this code, Fig. A.1 appears in the output.

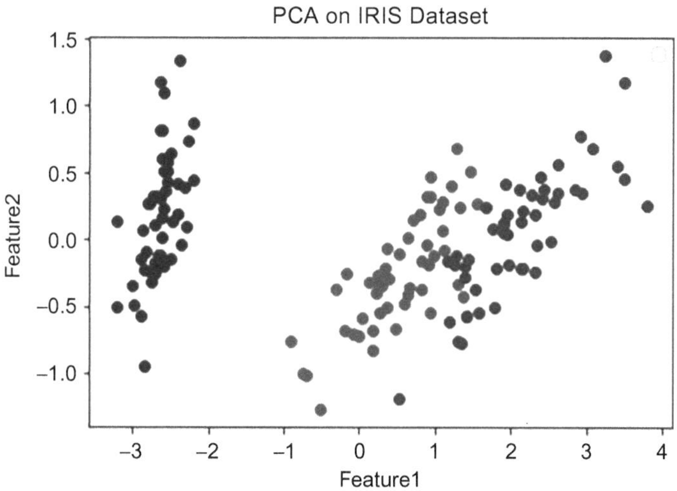

Fig. A.1 Result of PCA on IRIS dataset.

7.2 Partial least-squares

Partial least-squares (PLS) is a well-known technique of regression. This technique is used when the number of correlated independent variables is high. It should be noted that, in this case, the ordinary least square is failed. This situation is so common in chemometrics; hence, it is widely used in this research area. In this section, an example for PLS is given. In this example, physicochemical properties of protein tertiary structure dataset are utilized. This dataset is available in the machine learning repository. In the following, a python code is written to fit the PLS model on this dataset.

In the first step, the necessary libraries should be imported as follows:

```
import pandas as pd
import numpy as np
import matplotlib.pyplot as plt
from sklearn.preprocessing import scale
from sklearn import model_selection
from sklearn.model_selection import RepeatedKFold
from sklearn.cross_decomposition import PLSRegression
```

The "pandas" library is used to read the data since the file is saved in ".csv" format. Also, the "RepeatedKFold" class is imported since the model is evaluated by k-fold cross-validation and the "model_selection" class is imported to select the best model.

Next, the data should be read. To use this dataset, we need to download it and then read it. This part of code is given in the following:

```
data_url = "https://archive.ics.uci.edu/ml/machine-learning-
 databases/00265/CASP.csv"
raw_df = pd.read_csv(data_url, header=None)
data = raw_df.values[1::1, 1:10]
data = data.astype(np.float)
target = raw_df.values[1::1, 0]
target = target.astype(np.float)
```

Next, the PLS model fits into the data. In this case, there are nine predictors (namely, total surface area, non-polar exposed area, fractional area of exposed non-polar residue, fractional area of the exposed non-polar part of the residue, molecular mass-weighted exposed area, average deviation from the standard exposed area of residue, Euclidian distance, secondary structure penalty, and spatial distribution constraints) and one observable variable (i.e., size of the residue). In other words, the "data" variable is a $45,730 \times 9$ matrix and the "target" variable is a $45,730 \times 1$ matrix where 45,730 is a number of instances. As it is mentioned, in PLS, the predicted variables and the observable variables are projected into a new space; hence, the number of PLS components should be determined. To this end, for the different number of PLS components, the PLS model is fit, and then its MSE is computed. To do so, the following part of the code is used:

```
cv = RepeatedKFold(n_splits=5)
mse=[]
for i in np.arange(1, 9):
pls = PLSRegression(n_components=i)
score = -1*model_selection.cross_val_score(pls, scale(data),
target, cv=cv, scoring='neg_mean_squared_error').mean()
mse.append(score)
#plot number of PLS components vs. MSE
plt.plot(mse)
plt.xlabel('Number of PLS Components')
plt.ylabel('MSE')
```

The output of the above code is given in Fig. A.2.

As it is shown, when the number of PLS components is set to a value greater than three, the model gets the lowest mean-squared error.

7.3 Cluster analysis

In cluster analysis, the unlabeled data is divided into the number of groups such that samples in each group are similar to each other. In literature, there are many clustering algorithms (Mouton, Ferreira, & Helberg, 2020; Xu & Tian, 2015). In this section, an example for clustering is given and k-means is used as a clustering algorithm. Hence, an unlabeled dataset is needed to fit

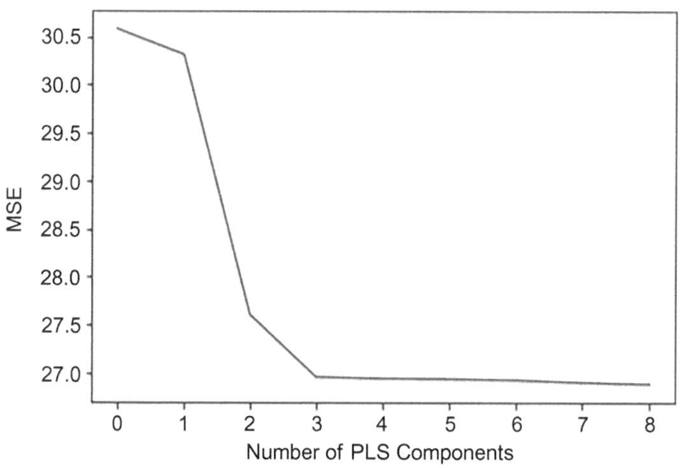

Fig. A.2 The effect of the number of PLS components vs. MSE on physicochemical properties of protein tertiary structure dataset.

the *k*-means model on the unlabeled dataset. In this section, the IRIS dataset is chosen to apply *k*_means on it. The reason for this selection is to visualize the clustering results and compare it with the real data. However, the class labels of samples are not utilized during the clustering algorithm. To evaluate the clustering algorithm in the real unlabeled dataset, the evaluation metrics like silhouette score, rand index, adjusted rand index, mutual information, and Davies-Bouldin index (Amigó, Gonzalo, Artiles, & Verdejo, 2009) are used. In the following, the python code to fit *k*-means on the IRIS dataset is given. Due to the visualization difficulty of high-dimensional data, first, data is projected into a new space with two dimensions and then *k*-means is applied to it. It should be noted that the step of dimensionality reduction is not necessary for all clustering algorithms. In this example, it is done due to the visualization goal.

Like the previous two examples, in the first step, the necessary libraries should be imported as follows:

```
import numpy as np
from sklearn import datasets
from sklearn.decomposition import PCA
import matplotlib.pyplot as plt
```

Then, data is read and PCA is applied on it to project data into a new space with two dimensions as follows:

```
from sklearn.datasets import load_iris
data = load_iris()
from sklearn.decomposition import PCA
pca = PCA(n_components=2)
data_reduced = pca.fit_transform(data.data)
```

Next, *k*_means is applied on "data_reduced" variable:

```
from sklearn.cluster import KMeans
kmeans = KMeans(n_clusters=2, random_state=0).fit(data_reduced)
```

As it is shown, first, by the "K_Means" class, an object from this class is constructed, and then by calling the "fit" function for this object, clustering is done. The "K_Means" class gets two arguments where the first one determines the number of clusters and the second one determines how the centroids initialization is done. Now, the learned clusters are visualized as follows:

```
idx = np.argwhere(kmeans.labels_==0)
plt.scatter(data_reduced[idx,0],data_reduced[idx,1],color='blue')
idx = np.argwhere(kmeans.labels_==1)
```

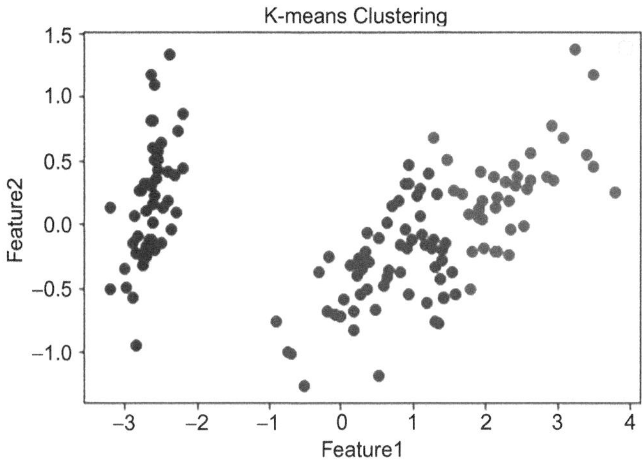

Fig. A.3 The obtained clusters of the *k*_means algorithm on the IRIS dataset.

```
plt.scatter(data_reduced[idx,0],data_reduced[idx,1],color='red')
idx = np.argwhere(kmeans.labels_==2)
plt.scatter(data_reduced[idx,0],data_reduced[idx,1],
 color='purple')
plt.gca().set(title="K-means Clusterig",
xlabel='Feature1', ylabel='Feature2')
plt.legend()
plt.show()
```

where the "kmeans.labels_" shows the assigned cluster numbers to each data. The result of the above code is shown in Fig. A.3.

By comparing Fig. A.1 with Fig. A.3, it is concluded that *k*_means clustering is done reliable on this dataset.

In Chapter 3, machine learning and pattern recognition methods in chemistry from multivariate and data-driven modeling are investigated. Also, many algorithms including linear regression, logistic regression, SVM, random forest, neural network (both for classification and for regression), and deep convolutional neural network (both for classification and for regression) are implemented on molecular property prediction by using DeepChem and scikit-learn (i.e., sklearn) packages. In the following, codes of these algorithms are given, and for each part of code, there are useful comments that explain the function of that part of code.

8. Linear regression code

The linear regression code is given in Table A.1.

Table A.1 Linear regression code.

```
import numpy as np import pandas as pd from sklearn import datasets, linear_model
from sklearn.metrics import mean_squared_error, r2_score import deepchem as dc
import matplotlib.pyplot as plt from sklearn.model_selection import train_test_split
from sklearn.decomposition import PCA # If you use Google Colab to run the code you
need to mount the drive # If you use your computer to run the code, you do not need these
commands from google.colab import drive drive.mount('/content/drive') # Read Data
data = pd.read_csv('/content/drive/My Drive/delaney-processed.csv')
smiles = data.loc[:, data.columns == 'smiles']
labels = data.loc[:, data.columns == 'ESOL predicted log solubility in mols per litre'] # use
DeepChem to extract ECFP features for each SMILES sequence
featurizer = dc.feat.CircularFingerprint(size=1024) XD=[] for i in range(len(smiles)):
ecfp = featurizer.featurize(smiles.loc[i]) ecfp = np.squeeze(ecfp) XD.append(ecfp) XD =
np.array(XD) labels = np.array(labels) # removes samples with NaN labels from the dataset
index = np.where(~np.isnan(labels)) labels = labels[index[0]] XD = XD[index[0],:] # split the
whole data into test and train sets
X_train, X_test, y_train, y_test = train_test_split(XD,labels, test_size=0.3) # creates a linear
regression model and then trains it using fit function # If you want Lasso regression you
should create model with this command # "regr = linear_model.Lasso(alpha=0.1)" or if you
want to create a Ridge # regression model use "regr = linear_model.Ridge(alpha=1.0)".
regr = linear_model.LinearRegression() regr.fit(X_train, y_train)
# Make predictions using the testing samples y_pred = regr.predict(X_train) # Calculate
two evaluation measures include MSE and r2_score and print # them
print(mean_squared_error(y_train, y_pred)) print(r2_score(y_train, y_pred)) # If you want
to plot the data, you should use dimensionality reduction # approach to project it into one
(2D space) or two dimensions (3D space). # To this end, PCA is used.
pca = PCA(n_components=1) X_train = pca.fit_transform(X_train) X_test =
```

Table A.1 Linear regression code—cont'd

pca.fit_transform(X_test) **plt.scatter(X_train,y_train,color='red',label="Training Samples")** **plt.scatter(X_test,y_test,color='purple',label="Test Samples") # To plot the predicted model in the projected space, you should need to # generate sample in the desired range and then predict the labels of # these samples by the learned model. Now, you can plot the predicted # model. xx = np.linspace(-3,3,1001) xx = np.expand_dims(xx,axis=-1) yy = regr.predict(xx) plt.plot(xx,yy,color='blue', label="Predicted Model") # Show the results plt.gca().set(title="Linear Regression on ESOL", xlabel='Feature', ylabel='Target Output') plt.legend() plt.show()**

9. Logistic regression code

The logistic regression code is given in Table A.2.

Table A.2 Logistic regression code.

```
import numpy as np import pandas as pd import deepchem as dc
from sklearn.model_selection import train_test_split
from sklearn.decomposition import PCA
from sklearn.linear_model import LogisticRegression
from sklearn.metrics import accuracy_score, roc_auc_score, f1_score
from sklearn.model_selection import KFold import matplotlib.pyplot as plt # If you use
Google Colab to run the code you need to mount the drive # If you use your computer to
run the code, you do not need these commands from google.colab import drive
drive.mount('/content/drive') # Read data
data = pd.read_csv('/content/drive/My Drive/tox21.csv')
smiles = data.loc[:, data.columns == 'smiles'] labels = data.loc[:, data.columns == 'NR-AR-
LBD'] # use DeepChem to extract ECFP features for each SMILES sequence
featurizer = dc.feat.CircularFingerprint(size=1024) XD=[] for i in range(len(smiles)):
ecfp = featurizer.featurize(smiles.loc[i]) ecfp = np.squeeze(ecfp) XD.append(ecfp) XD =
np.array(XD) labels = np.array(labels) # removes samples with NaN labels from the dataset
index = np.where(~np.isnan(labels)) labels = labels[index[0]] XD = XD[index[0],:] # To
evaluate the model k-fold cross-validations is used kf = KFold(n_splits=5) acc=[], acc_t=[]
f1=[], f1_t=[] roc=[], roc_t=[] for train_index, test_index in kf.split(XD):
print("TRAIN:", train_index, "TEST:", test_index)
X_train, X_test = XD[train_index], XD[test_index]
y_train, y_test = labels[train_index], labels[test_index] # Fit logistic regression models on
training samples
clf = LogisticRegression(random_state=0,class_weight='balanced').fit(X_train, y_train)#
y_pred = clf.predict(X_train) # Compute evaluation measures on training samples in each
fold acc.append(accuracy_score(y_train, y_pred)) f1.append(f1_score(y_train, y_pred))
roc.append(roc_auc_score(y_train, y_pred)) y_pred = clf.predict(X_test) # Compute
```

Table A.2 Logistic regression code—cont'd

evaluation measures on test samples in each fold **acc_t.append(accuracy_score(y_test, y_pred)) f1_t.append(f1_score(y_test, y_pred))** **roc_t.append(roc_auc_score(y_test, y_pred)) # Print the results on training samples** **print(np.mean(np.array(acc))) print(np.mean(np.array(f1))) print(np.mean(np.array(roc)))** **# Print the results on test samples print(np.mean(np.array(acc_t)))** **print(np.mean(np.array(f1_t))) print(np.mean(np.array(roc_t)))**

10. Support vector machine

The SVM is given in Table A.3.

Table A.3 Support vector machine.

```
import numpy as np import pandas as pd import deepchem as dc
from sklearn.linear_model import LogisticRegression
from sklearn.metrics import accuracy_score, roc_auc_score, f1_score
from sklearn.model_selection import KFold from sklearn.svm import SVC # If you use
Google Colab to run the code you need to mount the drive # If you use your computer to
run the code, you do not need these commands from google.colab import drive
drive.mount('/content/drive') # Read data
data = pd.read_csv('/content/drive/My Drive/tox21.csv')
smiles = data.loc[:, data.columns == 'smiles'] labels = data.loc[:, data.columns == 'NR-AR-
LBD'] # Use DeepChem to extract ECFP features for each SMILES sequence
featurizer = dc.feat.CircularFingerprint(size=1024) ecfp = featurizer.featurize(smiles)
XD=[] for i in range(len(smiles)): ecfp = featurizer.featurize(smiles.loc[i]) ecfp =
np.squeeze(ecfp) XD.append(ecfp) XD = np.array(XD) labels = np.array(labels) # Removes
samples with NaN labels from the dataset index = np.where(~np.isnan(labels)) labels =
labels[index[0]] XD = XD[index[0],:] # To evaluate the model, k-fold cross-validation is used
kf = KFold(n_splits=5) kf.get_n_splits(XD) acc = [], acc_t = [] f1 = [], f1_t = [] roc = [], roc_t
= [] for train_index, test_index in kf.split(XD):
X_train, X_test = XD[train_index], XD[test_index]
y_train, y_test = labels[train_index], labels[test_index] y_train = y_train + (-1)*(y_train==0)
y_test = y_test + (-1)*(y_test==0) # Fit SVM models on training samples. The kernel
parameter could be # set to another value. For more help, please see the documentation of #
the scikit-learn package.
clf = SVC(gamma='auto', kernel='rbf',class_weight='balanced').fit(X_train, y_train)
y_pred = clf.predict(X_train) # Compute evaluation measures on training samples in each
fold acc.append(accuracy_score(y_train, y_pred)) f1.append(f1_score(y_train, y_pred))
roc.append(roc_auc_score(y_train, y_pred)) y_pred = clf.predict(X_test) # Compute
```

Table A.3 Support vector machine—cont'd

evaluation measures on test samples in each fold acc_t.append(accuracy_score(y_test, y_pred)) f1_t.append(f1_score(y_test, y_pred)) roc_t.append(roc_auc_score(y_test, y_pred)) # Print the results on training samples print(np.mean(np.array(acc))) print(np.mean(np.array(f1))) print(np.mean(np.array(roc))) # Print the results on test samples print(np.mean(np.array(acc_t))) print(np.mean(np.array(f1_t))) print(np.mean(np.array(roc_t)))

11. Random forest code

The random forest code is given in Table A.4.

Table A.4 Random forest.

```
import numpy as np import pandas as pd import deepchem as dc
from sklearn.metrics import accuracy_score, roc_auc_score, f1_score
from sklearn.model_selection import KFold
from sklearn.ensemble import RandomForestClassifier # If you use Google Colab to run
the code you need to mount the drive # If you use your computer to run the code, you do not
need these commands from google.colab import drive drive.mount('/content/drive') # Read
data data = pd.read_csv('/content/drive/My Drive/tox21.csv')
smiles = data.loc[:, data.columns == 'smiles'] labels = data.loc[:, data.columns == 'NR-
AhR'] # could choose another task # Use DeepChem to extract ECFP features for each
SMILES sequence featurizer = dc.feat.CircularFingerprint(size=1024) XD = []
for i in range(len(smiles)): ecfp = featurizer.featurize(smiles.loc[i]) ecfp = np.squeeze(ecfp)
XD.append(ecfp) XD = np.array(XD) labels = np.array(labels) # Removes samples with
NaN labels from the dataset index = np.where(~np.isnan(labels)) labels = labels[index[0]]
XD = XD[index[0],:] # Use k-fold cross-validation to evaluate the model
kf = KFold(n_splits=5) kf.get_n_splits(XD) acc = [], acc_t = [] f1 = [], f1_t = [] roc = [], roc_t
= [] for train_index, test_index in kf.split(XD):
print("TRAIN:", train_index, "TEST:", test_index)
X_train, X_test = XD[train_index], XD[test_index]
y_train, y_test = labels[train_index], labels[test_index] # Fit the random forest model on
training samples. The "n_estimators", # "criterion", and "max_depth" could be set to
another value. For # more help, please see the documentation of the scikit-learn package.
clf = RandomForestClassifier(n_estimators=100,criterion='gini',class_weight='balanced',m
ax_depth=4).fit(X_train, y_train)# y_pred = clf.predict(X_train) # Compute evaluation
measures on training samples in each fold acc.append(accuracy_score(y_train, y_pred))
f1.append(f1_score(y_train, y_pred)) roc.append(roc_auc_score(y_train, y_pred))
y_pred = clf.predict(X_test) # Compute evaluation measures on test samples in each fold
acc_t.append(accuracy_score(y_test, y_pred)) f1_t.append(f1_score(y_test, y_pred))
```

Table A.4 Random forest—cont'd

roc_t.append(roc_auc_score(y_test, y_pred)) # Print the results on training samples print(np.mean(np.array(acc))) print(np.mean(np.array(f1))) print(np.mean(np.array(roc))) # Print the results on training samples print(np.mean(np.array(acc_t))) print(np.mean(np.array(f1_t))) print(np.mean(np.array(roc_t)))

12. Deep learning for regression task

The deep learning for the regression task code is given in Table A.5.

Table A.5 Deep learning for regression task.

```
import numpy as np import pandas as pd import keras from keras.models import Model
from keras.preprocessing import sequence
from keras.models import Sequential, load_model
from keras.layers import Dense, Dropout, Activation from keras.layers import Embedding
from keras.models import Model
from keras.callbacks import ModelCheckpoint, EarlyStopping
from keras import optimizers, layers from tensorflow.keras import regularizers
from sklearn.model_selection import train_test_split
from sklearn.metrics import mean_squared_error, r2_score # If you use Google Colab to
run the code you need to mount the drive # If you use your computer to run the code, you
do not need these commands from google.colab import drive drive.mount('/content/drive') #
Read data data = pd.read_csv('/content/drive/My Drive/delaney-processed.csv')
smiles = data.loc[:, data.columns == 'smiles']
labels = data.loc[:, data.columns == 'ESOL predicted log solubility in mols per litre'] #
Show each element of SMILES sequence with an index (integer number)
smiles_dict = {"#": 29, "%": 30, ")": 31, "(": 1, "+": 32, "-": 33, "/": 34,
".": 2, "1": 35, "0": 3, "3": 36, "2": 4, "5": 37, "4": 5, "7": 38, "6": 6, "9": 39, "8": 7,
"=": 40, "A": 41, "@": 8, "C": 42, "B": 9, "E": 43, "D":10, "G": 44, "F": 11,
"I": 45, "H": 12, "K": 46, "M": 47, "L": 13, "O": 48, "N": 14, "P": 15, "S": 49,
"R": 16, "U": 50, "T": 17, "W": 51,"V": 18, "Y": 52,
"[": 53, "Z": 19, "]": 54, "\": 20, "a": 55,"c": 56, "b": 21, "e": 57, "d": 22, "g": 58,
"f": 23, "i": 59, "h": 24, "m": 60, "l": 25, "o": 61, "n": 26, "s": 62,
"r": 27, "u": 63, "t": 28, "y": 64, " ":65, ":": 66, ",":67, "p":68, "j": 69}
def label_smiles(line, MAX_SMI_LEN, smi_ch_ind): X = np.zeros(MAX_SMI_LEN)
for i, ch in enumerate(line[:MAX_SMI_LEN]): if ch != '\n': X[i] = smi_ch_ind[ch] return X
XD = [] for i in range(len(smiles)):
XD.append(label_smiles(str(smiles.loc[i]), 100, smiles_dict)) # Split the whole data into train
```

Table A.5 Deep learning for regression task—cont'd

and test set X_train, X_test, y_train, y_test = train_test_split(XD,labels, test_size=0.3) # Define the architecture of the model XDinput = Input(shape=(100,)) FC1 = Dense(1024, activation='relu')(XDinput) FC2 = Dropout(0.1)(FC1) FC2 = Dense(1024, activation='relu')(FC2) FC2 = Dropout(0.1)(FC2) FC2 = Dense(512, activation='relu')(FC2) predictions = Dense(1,activity_regularizer=regularizers.l1(1e-5))(FC2) # Create the model interactionModel = Model(inputs=XDinput, outputs=predictions) # Print the architecture of the model in the output interactionModel.summary() # Compile the model. Choose loss function, optimizer algorithm and metrics interactionModel.compile(optimizer='sgd', loss='mse', metrics=['mse']) # Prepare the training data to feed into the model XD = np.array(X_train) labels = np.array(y_train) # Removes samples with NaN labels from the training samples index = np.where(~np.isnan(labels)) labels = labels[index[0]] XD = XD[index[0],:] # Train the model interactionModel.fit(XD,labels, batch_size=256,epochs = 100) # Predict on the training samples y_pred = interactionModel.predict(np.array(XD)) # Print the evaluation measures on the training samples print(mean_squared_error(labels, y_pred)) print(r2_score(labels, y_pred)) # Prepare the test data to feed into the model XD = np.array(X_test) labels = np.array(y_test) # Removes samples with NaN labels from the test samples index = np.where(~np.isnan(labels)) labels = labels[index[0]] XD = XD[index[0],:] # Predict on the test samples y_pred = interactionModel.predict(np.array(XD)) # Print the evaluation measures on the test samples print(mean_squared_error(labels, y_pred)) print(r2_score(labels, y_pred))

13. Neural network on classification task

The neural network on classification task code is given in Table A.6.

Table A.6 Neural network on classification task.

```
import numpy as np import pandas as pd import keras from keras.models import Model
from keras.preprocessing import sequence
from keras.models import Sequential, load_model
from keras.layers import Dense, Dropout, Activation from keras.layers import Embedding
from keras.layers import Conv1D, MaxPooling1D from keras.layers import Conv2D, GRU
from keras.layers import Input, Embedding, Dense from keras.models import Model
from keras import optimizers, layers from tensorflow.keras import regularizers
from sklearn.model_selection import train_test_split
from sklearn.model_selection import KFold # If you use Google Colab to run the code you
need to mount the drive # If you use your computer to run the code, you do not need these
commands from google.colab import drive drive.mount('/content/drive') # Read data
data = pd.read_csv('/content/drive/My Drive/tox21.csv')
smiles = data.loc[:, data.columns == 'smiles'] labels = data.loc[:, data.columns == 'NR-AR-
LBD'] # Show each element of SMILES sequence with an index (integer number)
smiles_dict = {"#": 29, "%": 30, ")": 31, "(": 1, "+": 32, "-": 33, "/": 34,
".": 2, "1": 35, "0": 3, "3": 36, "2": 4, "5": 37, "4": 5, "7": 38, "6": 6, "9": 39, "8": 7,
"=": 40, "A": 41, "@": 8, "C": 42, "B": 9, "E": 43, "D":10, "G": 44, "F": 11,
"I": 45, "H": 12, "K": 46, "M": 47, "L": 13, "O": 48, "N": 14, "P": 15, "S": 49,
"R": 16, "U": 50, "T": 17, "W": 51,"V": 18, "Y": 52,
"[": 53, "Z": 19, "]": 54, "\": 20, "a": 55,"c": 56, "b": 21, "e": 57, "d": 22, "g": 58,
"f": 23, "i": 59, "h": 24, "m": 60, "l": 25, "o": 61, "n": 26, "s": 62,
"r": 27, "u": 63, "t": 28, "y": 64, " ":65, ":": 66, ",":67, "p":68, "j": 69}
def label_smiles(line, MAX_SMI_LEN, smi_ch_ind): X = np.zeros(MAX_SMI_LEN)
for i, ch in enumerate(line[:MAX_SMI_LEN]): if ch!='\n': X[i] = smi_ch_ind[ch] return X
XD = [] for i in range(len(smiles)):
XD.append(label_smiles(str(smiles.loc[i]), 100, smiles_dict)) XD=np.array(XD)
```

Table A.6 Neural network on classification task—cont'd

```
labels=np.array(labels) index=np.where(~np.isnan(labels)) labels=labels[index[0]]
XD=XD[index[0],:] # Define the architecture of the model XDinput = Input(shape=(100,))
FC1 = Dense(1024, activation='relu')(XDinput) FC2 = Dropout(0.1)(FC1)
FC2 = Dense(1024, activation='relu')(FC1) FC2 = Dropout(0.1)(FC2)
FC2 = Dense(512, activation='relu')(FC2) predictions = Dense(1, activation='sigmoid')(FC2)
# Create the model interactionModel = Model(inputs=XDinput, outputs=predictions) #
Print the architecture of the model in the output interactionModel.summary() # Define the
evaluation metrics to evaluate the model METRICS = [
keras.metrics.BinaryAccuracy(name='accuracy'),
keras.metrics.Precision(name='precision'), keras.metrics.Recall(name='recall'),
keras.metrics.AUC(name='auc'), ] # Compile the model. Choose loss function, optimizer
algorithm and metrics
interactionModel.compile(optimizer='adam', loss='binary_crossentropy', metrics=[METRI
CS]) # Use k-fold cross-validation to evaluate the model kf = KFold(n_splits=5)
train_results = [] test_results = [] for train_index, test_index in kf.split(XD):
print("TRAIN:", train_index, "TEST:", test_index)
X_train, X_test = XD[train_index], XD[test_index]
y_train, y_test = labels[train_index], labels[test_index] X_train = np.array(X_train) y_train
= np.array(y_train) # Removes samples with NaN labels from the training samples index =
np.where(~np.isnan(y_train)) y_train = y_train[index[0]] X_train = X_train[index[0],:] #
Train the defined neural network model on training samples
interactionModel.fit(X_train,y_train, batch_size=256,epochs = 100, class_weight={0: 0.1, 1:
0.9}) train_results.append(interactionModel.evaluate(X_train,y_train)) X_test =
np.array(X_test) y_test = np.array(y_test) # Removes samples with NaN labels from the test
samples index = np.where(~np.isnan(y_test)) y_test = y_test[index[0]] X_test =
X_test[index[0],:] test_results.append(interactionModel.evaluate(X_test,y_test)) # Print the
evaluation metrics on the training samples print(np.mean(train_results,axis=0)) # Print the
evaluation metrics on the test samples print(np.mean(test_results,axis=0))
```

14. Convolutional neural network

The convolutional neural network code is given in Table A.7.

Table A.7 Convolutional neural network.

```
import numpy as np import pandas as pd import keras from keras.models import Model
from keras.preprocessing import sequence
from keras.models import Sequential, load_model
from keras.layers import Dense, Dropout, Activation from keras.layers import Embedding
from keras.layers import Conv1D, MaxPooling1D from keras.layers import Conv2D, GRU
from keras.layers import Input, Embedding, Dense, Flatten
from keras.models import Model
from keras.callbacks import ModelCheckpoint, EarlyStopping
from keras import optimizers, layers from tensorflow.keras.utils import to_categorical
from tensorflow.keras import regularizers
from sklearn.model_selection import train_test_split
from sklearn.model_selection import KFold # If you use Google Colab to run the code you
need to mount the drive # If you use your computer to run the code, you do not need these
commands from google.colab import drive drive.mount('/content/drive') # Read data
data = pd.read_csv('/content/drive/My Drive/tox21.csv')
smiles = data.loc[:, data.columns == 'smiles'] labels = data.loc[:, data.columns == 'SR-
ARE'] # Show each element of SMILES sequence with an index (integer number)
smiles_dict = {"#": 29, "%": 30, ")": 31, "(": 1, "+": 32, "-": 33, "/": 34,
".": 2, "1": 35, "0": 3, "3": 36, "2": 4, "5": 37, "4": 5, "7": 38, "6": 6, "9": 39, "8": 7,
"=": 40, "A": 41, "@": 8, "C": 42, "B": 9, "E": 43, "D":10, "G": 44, "F": 11,
"I": 45, "H": 12, "K": 46, "M": 47, "L": 13, "O": 48, "N": 14, "P": 15, "S": 49,
"R": 16, "U": 50, "T": 17, "W": 51,"V": 18, "Y": 52,
"[": 53, "Z": 19, "]": 54, "\": 20, "a": 55,"c": 56, "b": 21, "e": 57, "d": 22, "g": 58,
"f": 23, "i": 59, "h": 24, "m": 60, "l": 25, "o": 61, "n": 26, "s": 62,
"r": 27, "u": 63, "t": 28, "y": 64, " ":65, ":": 66, ",":67, "p":68, "j": 69}
def label_smiles(line, MAX_SMI_LEN, smi_ch_ind): X = np.zeros(MAX_SMI_LEN)
for i, ch in enumerate(line[:MAX_SMI_LEN]): # x, smi_ch_ind, y if ch != '\n':
X[i] = smi_ch_ind[ch] return X XD = [] for i in range(len(smiles)):
XD.append(label_smiles(str(smiles.loc[i]), 100, smiles_dict)) # Define the architecture of the
model XDinput = Input(shape=(100,70)) # Feature extraction layers
```

Table A.7 Convolutional neural network—cont'd

```
encode_smiles= Conv1D(filters=64, kernel_size=2, activation='relu', padding='valid',
strides=1)(XDinput) encode_smiles = Conv1D(filters=64*2, kernel_size=4,
activation='relu', padding='valid', strides=1)(encode_smiles)
encode_smiles = Conv1D(filters=64*3, kernel_size=4, activation='relu', padding='valid',
strides=1)(encode_smiles) encode_smiles = GlobalMaxPooling1D()(encode_smiles)
encode_smiles = Flatten()(encode_smiles) # Fully connected layers (task prediction layers)
FC1 = Dense(1024, activation='relu')(encode_smiles) FC2 = Dropout(0.1)(FC1)
FC2 = Dense(1024, activation='relu')(FC1) FC2 = Dropout(0.1)(FC2)
FC3 = Dense(512, activation='relu')(FC2) predictions = Dense(1, activation='sigmoid')(FC3)
# Create the model interactionModel = Model(inputs=XDinput, outputs=predictions) #
Print the architecture of the model in output interactionModel.summary() # Define the
metrics to evaluate the model METRICS = [
keras.metrics.BinaryAccuracy(name='accuracy'),
keras.metrics.Precision(name='precision'), keras.metrics.Recall(name='recall'),
keras.metrics.AUC(name='auc') ] # compiles the model. Choose loss function, optimizer
algorithm and # metrics.
interactionModel.compile(optimizer='adam', loss='binary_crossentropy', metrics=[METRI
CS]) # Use k-fold cross-validation to evaluate the model kf = KFold(n_splits=5)
train_results = [] test_results = [] for train_index, test_index in kf.split(XD):
X_train, X_test = XD[train_index], XD[test_index]
y_train, y_test = labels[train_index], labels[test_index] X_train = np.array(X_train) y_train
= np.array(y_train) # One-hot encoding of the training samples X_train =
to_categorical(X_train,70) # Removes samples with NaN labels from the training samples
index = np.where(~np.isnan(y_train)) y_train = y_train[index[0]] X_train =
X_train[index[0],:] # Train the model using the training samples
interactionModel.fit(X_train,y_train, batch_size=256,epochs = 100, class_weight={0: 0.1, 1:
0.9}) train_results.append(interactionModel.evaluate(X_train,y_train)) X_test =
np.array(X_test) y_test = np.array(y_test) # One-hot encoding of the test samples X_test =
```

Continued

Table A.7 Convolutional neural network—cont'd

to_categorical(X_test,70) # Removes samples with NaN labels from the test samples index = np.where(~np.isnan(y_test)) y_test = y_test[index[0]] X_test = X_test[index[0],:] test_results.append(interactionModel.evaluate(X_test,y_test)) # Print the results on the training samples print(np.mean(train_results,axis=0)) # Print the results on the test samples print(np.mean(test_results,axis=0))

References

Amigó, E., Gonzalo, J., Artiles, J., & Verdejo, F. (2009). A comparison of extrinsic clustering evaluation metrics based on formal constraints. *Information Retrieval*, *12*(4), 461–486.

Camp, C. H., Jr. (2019). PyMCR: A python library for multivariatecurve resolution analysis with alternating regression (MCR-AR). *Journal of Research of the National Institute of Standards and Technology*, *124*, 1–10.

Cao, D. S., Liang, Y. Z., Yan, J., Tan, G. S., Xu, Q. S., & Liu, S. (2013). PyDPI: Freely available python package for chemoinformatics, bioinformatics, and chemogenomics studies. *ACS Journal of Chemical Information and Modeling*, *53*(11), 3086–3096.

Cao, D. S., Xu, Q. S., Hu, Q. N., & Liang, Y. Z. (2013). ChemoPy: Freely available python package for computational biology and chemoinformatics. *Bioinformatics*, *29*(8), 1092–1094.

Mouton, J. P., Ferreira, M., & Helberg, A. S. (2020). A comparison of clustering algorithms for automatic modulation classification. *Expert Systems with Applications*, *151*, 113317.

Salgado, P. (2021). *PEP 619—Python 3.10 release schedule.*

Stewart, J. J. (1990). MOPAC: A semiempirical molecular orbital program. *Journal of Computer-Aided Molecular Design*, *4*(1), 1–103.

Van Rossum, G., & Drake, F. L., Jr. (1995a). *Python reference manual.* Amsterdam: Centrum voor Wiskunde en Informatica.

Van Rossum, G., & Drake, F. L., Jr. (1995b). *Python tutorial. Vol. 620.* Amsterdam: Centrum voor Wiskunde en Informatica.

Xu, D., & Tian, Y. (2015). A comprehensive survey of clustering algorithms. *Annals of Data Science*, *2*(2), 165–193.

Index

Note: Page numbers followed by *f* indicate figures and *t* indicate tables.

H

K

L

M

N

O

P

R

S

T

U

V

W

Kolibri 360 GmbH
Ferdinand-Jühlke-Straße 7
99095 Erfurt, Deutschland
produktsicherheit@kolibri360.de